CHEMISTRY OF
VEGETABLE
TANNINS

Dedicated to Prof. R. D. Haworth, F.R.S., Firth Professor and Head of the Department of Chemistry at the University of Sheffield from 1939 to 1963.

CHEMISTRY OF
VEGETABLE
TANNINS

by

E. HASLAM

*Department of Chemistry
The University,
Sheffield, England*

1966

ACADEMIC PRESS
LONDON and NEW YORK

ACADEMIC PRESS INC. (LONDON) LTD
Berkeley Square House
Berkeley Square
London, W.1

U.S. Edition published by

ACADEMIC PRESS INC.
111 Fifth Avenue
New York, New York 10003

Library of Congress Catalog Card Number: 65–27887

Printed in Great Britain by
William Clowes and Sons, Limited, London and Beccles

Preface

The vegetable tannins occupy a unique part of the borderland between chemistry and botany and their use in the art of leather manufacture has led to an association in the original literature not only with these scientific disciplines but also with the various skills, practices and trades employed by man since very early times. The chemistry of the vegetable tannins has remained something of an enigma; although Emil Fischer was able to make outstanding contributions to the study of the hydrolysable tannins in the early years of this century, later work, particularly with the condensed or non-hydrolysable tannins, mainly served to illustrate the complexity of problems in this field and until comparatively recently the subject was in a very disorderly state. The advent of new techniques for the analysis and separation of water soluble materials has led within a very short period of time (as in related fields of organic chemistry) to substantial progress in the chemistry of these substances. Whilst detailed knowledge of the diverse structures of many of the hydrolysable tannins has been realized the same cannot yet be said for the condensed tannins. Advances have nevertheless been made towards the identification of their precursors and the probable pathways for their formation in plant tissues, and it is hoped that with the generally increased interest in this area of plant biochemistry fundamental progress will not be too long awaited.

The objects of this text are twofold. Firstly, to provide a comprehensive survey of the recent achievements and the many problems which remain in the chemistry of the vegetable tannins and to define more precisely its relationship with other fields of natural products chemistry. Secondly, to give a synopsis of the methods which have had the most successful application in recent studies and which, in the author's view, will find particular use in further investigations. In the process it is hoped that others will derive an interest in the subject and engage in future work on the chemistry of plant polyphenols.

It is a pleasure to record the author's gratitude to several colleagues and friends who advised on the nature of the manuscript, and particularly to Dr. T. Swain whose comments were not only interesting but useful.

University of Sheffield   E. HASLAM
November 1965

Contents

CHAPTER 5. The Biosynthesis of Plant Polyphenols

CHAPTER 1

The Scope of Vegetable Tannin Chemistry

I. INTRODUCTION

Leather making is a craft of great antiquity and records exist relating to its operation in Mediterranean regions around 1500 B.C. The word tanner to describe a person who pursued this trade has probably been used for a similar period of time but it is doubtful if the corresponding term tannin to denote the substances responsible for the conversion of raw animal skins into leather was in common usage much before the end of the eighteenth century. Vegetable tannin chemistry can indeed be said to have its origins at this time, for only then was it generally recognized that the tanning process involved a combination of substances—tannins—in the plant extract with the animal skins and was not merely a vaguely defined physical change in which the astringent tanning liquors caused the skins to harden and shrink. As befits an industry with this historical background the tanner's art has been gained by countless years of observation by hand and eye and frequently taste, and is one which in some respects has been influenced surprisingly little by the scientific revolution of the twentieth century.

The process of tannage consists essentially in the production from raw animal hides and skins of materials, normally referred to as leathers, which have a greater stability to water, bacteria, heat and abrasion and have as a consequence a wide range of domestic and industrial applications. The type of leather produced is to a large extent dependent on the origin and previous treatment of the skin but some control may also be effected by the use of different tanning materials (inorganic, vegetable or synthetic) and by variation of the conditions (temperature, pH and duration) employed during tannage. Ox hides give leathers with a high tensile strength used in the production of leather belting, calf hides are tanned to give the flexible upholstery

1*

leather and clothing and ornamental leathers are generally obtained from sheepskin, goatskin and the skins of reptiles. Despite many years of patient research much of the scientific background to the tanner's art remains, at best, only partially understood and although a rational explanation is possible in qualitative terms for many of the traditional forms of tanning there is little doubt that considerably more remains to be learnt scientifically about these processes.

The vegetable tannins are polyphenols with a molecular weight in the range 500–3000 and as their name suggests are tanning materials from a plant source. Although they have several other industrial and technological applications (such as in the manufacture of inks and plastics, the preservation of fish nets, in oil-well drilling and as mordants in dyeing) any definition of the scope of vegetable tannin chemistry must recognize their intimate relationship to leather production. In the simplest terms the chemistry of leather manufacture is seen as the interaction of inorganic, natural or synthetic tannins with the collagen fibres of the corium of skin and in order to discuss this reaction in more detail some elementary appreciation of the structure of the protein collagen is therefore necessary. Considerable refinements, particularly of the physical methods for the analysis of protein structures, have enabled a fairly detailed model of collagen fibres to be built up and the evidence leading to the proposed structure is briefly reviewed below.

II. The Structure of Collagen

The protein collagen which has a fibrous appearance at all levels of optical resolution available[1, 2] forms a variety of patterns in the various tissues of the animal body. In cow-hide the fibres branch and anastomose, in tendon they are all orientated parallel to the long axis of the tendon and in tissues such as the cornea they are arranged in fine laminated sheets. In the light microscope the fibres in tendon have diameters of the order of 100–200 μ and some four or five times less in skin. In the electron microscope, fibrils—subdivisions of the fibre—are seen whose diameters show a continuous range varying from perhaps a thousand angstroms to the resolution of the instrument. Collagen preparations which have been stained with uranyl salts or phosphotungstic acid and are then examined under the electron microscope show a series of dark bands with light unstained regions between them with the periodicity of the band repeat along the fibre axis approximating to 640 Å. The low angle X-ray diffraction pattern of collagen also reveals a well-defined repeating unit of 640 Å along the fibre axis and this and the characteristic electron micrographic patterns have been interpreted in terms of the same gross structural feature in the collagen structure which is discussed in more detail below. All collagen fibres also give a distinct wide angle X-ray diffraction pattern which points to the existence of a high degree of order in certain parts of the fibre. The prominent spacing of 2·86 Å along the fibre axis in the wide angle X-ray diffraction, combined with the chemical analysis of the protein has been interpreted by Crick and others[3, 4] in terms of a unique polypeptide chain configuration in the ordered parts of the collagen molecule.

Chemically collagen from a number of animal sources is distinguished by its unusually high content of glycine (1), proline (2; R = H) and hydroxyproline (2; R = OH) which together account for over 50% of the amino acid content of the protein[5]. Sufficient information is now available following the isolation and identification of large numbers of di-, tri- and tetrapeptides from acid and enzymic hydrolysates of collagen to establish the essential features of the primary structural patterns which predominate in its polypeptide chains. Glycine is widely distributed amongst the different oligopeptides derived from collagen hydrolysates and this supports the original suggestion of Astbury[6] that glycine occurs as every third residue in the polypeptide chain, although it is also apparent from the structure of some of the peptides that this arrangement is not absolute. The imino acids proline and hydroxyproline are located[7] in these peptides in a regular manner— thus the majority occur as glycine·proline and hydroxyproline·glycine respectively—and since they are also present in almost equal amounts it has been inferred that amino acid sequences of the type glycine·proline·hydroxyproline·glycine form important units in the primary structure of collagen. This hypothesis has also been supported in particular by the work of Kroner[8] and his collaborators who isolated and identified this tetrapeptide in significant amounts from collagen hydrolysates.

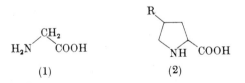

(1)   (2)

Interpretations of the distinctive wide angle X-ray diffraction pattern of the collagens in terms of a molecular structure are based essentially on the accommodation of the repeating sequence of amino acids glycine·proline (or any other amino acid)·hydroxyproline (or any other amino acid) which chemical degradations suggest are widely distributed in the protein[9]. Ramachandran and Kartha[4] were the first to propose a triple helix structure for collagen consisting of three polypeptide chains slowly twisting around one another and linked by hydrogen bonds. Rich and Crick[3] later criticized fine structural details of this model on stereochemical grounds and proposed two alternative structures, collagen I and II, based however on a similar triple helix model. The configurations of the polypeptide chains in these structures were closely related to that of the model compound polyglycine. Construction of accurate molecular models showed that collagen II was much easier to build, gave more acceptable values of bond dimensions, angles and interatomic distances and readily accommodated the critical amino acid sequence glycine·proline·hydroxyproline. Collagen II was therefore assumed to be an accurate representation of the collagen molecule and a model of this structure is shown in Fig. 1. The three polypeptide chains in collagen II are twisted about each other in a gradual right-hand helix with a screw axis of 108° and a translation of 2·86 Å. Each chain is separated from the other two by a distance of 5 Å and the whole structure is held together by hydrogen bonding between every third —NH— group on the backbone of one chain and every third $\rangle$C$=$O group of a neighbour. The hydroxyl groups of the hydroxyproline residues point radially outwards from the structure and may serve to link groups of collagen chains together in the fibrils by hydrogen bonding. The molecular model (Fig. 1), which resembles in many respects a triple stranded electric flex, appears as a cylinder with a deep helical groove (due to the glycine residues with no carbon side chain), wound around it. The corresponding helical ridge is due to the side chains of the other amino acids and Rich and Crick suggested that collagen molecules when laid down side by side in the fibre could be envisaged as "screwed" together with the ridge of one molecule lying next to the groove of its neighbour. This was suggested as a further possible factor in the stabilization of the collagen molecule in the intact fibre.

Although the construction of skeletal models indicated that collagen II was superior to all other models it is possible that different parts of the collagen molecule may have different configurations of the basic triple helix structure. Thus certain regions may be laid down in the alternative collagen I form, which with small deformation is able to accommodate the determining amino acid sequence glycine·proline·hydroxyproline, or in a form similar to the revised triple helical structures of Ramachandran[10]. However, no information is available to assess the importance of alternative polypeptide chain configurations in the collagen structure and it is assumed here that the collagen II model is a substantially accurate representation of the ordered parts of the protein molecule.

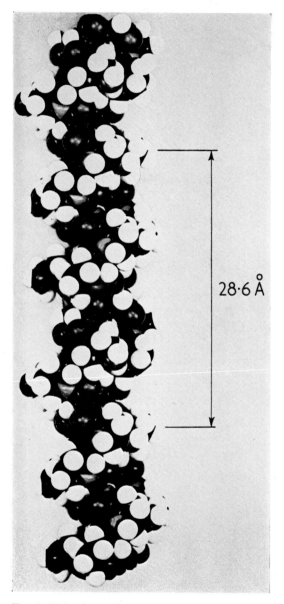

28·6 Å

Fig. 1. Molecular model of collagen (Rich and Crick).

In collagen hydrolysates considerable quantities (approximately 20%) of various amino acids (such as glutamic acid, arginine and asparagine), containing acidic, amide and basic side chains are present[1]. If these are assumed to be incorporated into the collagen II molecular structure by replacement of proline and hydroxyproline residues then their longer side chains containing polar groups might be expected to change the shape and nature of the helical ridge, and thus to influence the close packing and stability of the intermolecular links in the fibre. That this is in fact the case follows from the work of Grassmann[11, 12] and Schmitt[13-15] and their colleagues who respectively deduced a more detailed picture of the primary structure of collagen and a rational explanation of the way in which the collagen molecules pack together in the fibrils. Grassmann and his collaborators isolated over fifty peptides, containing from three to more than a hundred amino acid residues, from tryptic digests of collagen. Analysis of these peptides showed the proline and hydroxyproline rich ones to be deficient in the amino acids with polar side chains and conversely those rich in this type of acid contained little if any of the imino acids. These results Grassmann has suggested support the original suggestion of Bear who proposed[2] that the "band–interband" repeat observed in electron micrographs of stained collagen preparations were due to regular alternation in the fibrils of groups of amino acids, the "bands" containing high concentrations of the long polar amino acid side chains, and the "interbands" mainly the smaller non-polar residues. Bear also suggested that the "interbands" were the ordered crystalline portions which give rise to the distinct low angle X-ray pattern and that the bulky polar amino acids present in the "bands" prevent proper close packing of the collagen molecules at these parts of the fibrils and result in its amorphous and readily stained regions. Independent confirmation of this concept of fibrillar structure comes from the work of Schmitt and his colleagues[13-15] with soluble collagen—prepared by solution of collagen in dilute acid. Using minor but critical variations in the solvent environment Schmitt was able to reprecipitate collagen from solution in at least five different fibrous modifications whose structures were interpreted in terms of a hypothetical polarized collagen molecule—"tropocollagen"—with dimensions of the order of 2600 Å in length and 15–20 Å in diameter. The observed fibrous forms were postulated as being produced by different alignments of the "tropocollagen" precursor (Fig. 2). Arrangement of the molecules in parallel with the ends in register gives the segment long spacing (s.l.s.) pattern in which all the like features of a series of collagen molecules are in accurate transverse register (Fig. 2(b)). In native collagen the tropocollagen molecules

originally suggested by Bear[2] and which the detailed work of Grassmann, Schmitt, Doty and their colleagues supports is shown in Fig. 3.
The close and regular packing of the collagen molecules gives rise to the
crystalline portions of the fibril; where this intimate alignment is dis-
torted by the bulky polar amino acid side chains the fibril has an
amorphous and less ordered structure. The model is similar in some
respect to the structure of deoxyribonucleic acid, although the type of
feature which is complementary in one collagen chain to that in another
and which permits the polypeptide chains to be held together at these
points, is not clearly understood. Schmitt and his collaborators have
suggested that these discrete interactions arise from specific arrangements of the polar amino acids along the polypeptide chains. When the

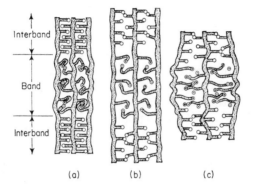

FIG. 3. Fibrillar structure of collagen[2]. (a) Dry fibril; (b) fibril swelling in water;
(c) fibril swelling in acid.

fibril imbibes water the molecular structure is seen as becoming less
ordered in all regions (Fig. 3(b)). When, however, the fibril is swollen by
acids, preferential separation of the collagen chains is suggested as
occurring mainly in the amorphous regions containing the polar amino
acid side chains (Fig. 3(c)).

III. The Mechanism of Vegetable Tannage

Most workers agree that the essence of vegetable tanning is to protect
the molecular form of the collagen fibres and that this is achieved by
packing the amorphous regions of the fibrillar structure with tannins[19, 20]. The crystalline parts do not require tanning, since the
orderly arrangement of the molecular chains in these regions offers
protection against the penetration of water and attack by bacteria.

The first stage of tanning is envisaged as a combination of the hydroxyl groups present in the vegetable tannins with "active sites" in the exposed amorphous regions of the collagen structure, this being followed by further bonding of the fixed tannins with other tannin molecules until all the available space between the collagen chains is filled. During tannage the accessibility of the "active sites" to the tannin molecules is further increased by osmotic swelling of the fibrillar structure in the aqueous acidic environment (Figs. 3(b) and (c)). Evidence from a number of sources indicates that this initial process is complete and the fibre tanned when the collagen has adsorbed approximately half its weight of vegetable tannins. It is of interest to note that the initial stage in chrome tannage—using basic chromium salts as the tanning agent—is complete when the skin has taken up as little as 3% of the tannin and stabilization of the fibres in this case is thought to be achieved in a completely different manner by the formation of specific non-ionizable links between the polypeptide chains[19, 20]. Further degrees of vegetable tannage are possible by means of heavy leather tanning which may take up to two years to complete. The operation probably proceeds by additional tannin molecules passing slowly into the crystalline portions of the fibrils and is assisted by the use of low pH tanning liquors and elevated temperatures. Chemical analysis[21] has shown the vegetable tannins to be generally amorphous substances characterized by an accumulation of aliphatic and phenolic hydroxyl, and in some cases carboxyl, groups. On the basis of a similar qualitative view of vegetable tannage to that outlined here, White[21] has further suggested that for a substance to act as a vegetable tannin it must be able to penetrate the interfibrillar regions of collagen fibres but at the same time sufficiently large to cross-link chains at more than one point. Polyphenols with a molecular weight range of 500–3000 were postulated by White as most satisfactorily fulfilling these requirements.

A question which is still debated is the nature of the "active sites" on the collagen chains and the nature of the reaction between these sites and the tannin molecules which takes place during tannage. A number of findings favour the view that these "active sites" are the peptide bonds which are not internally bonded to other groups and are thus able to form similar hydrogen bonded structures with the phenolic groups of the tannin. It is also probable that other groups such as amide and amine residues present in polar amino acid side chains are involved in the fixation of vegetable tannins on a limited scale. The evidence favouring the peptide bonds as the principal "active sites" in the collagen molecule during tannage is derived from a number of sources. Thus Grassmann[22] has shown that water soluble condensation products of urea and formaldehyde—such as ($-CH_2-NH-CO-NH-CH_2-)_n$ which contain the peptide bond as the sole reactive grouping—precipitate vegetable tannins quantitatively. Gustavson[23, 24] and Batzer and Weissenberger[25] have similarly provided direct evidence for the participation of the peptide bond of collagen in the vegetable tanning process by their model experiments with synthetic polyamides and esterified forms of collagen. Earlier attempts to use polyamides, such as nylon in the unorientated state, as substrates for tanning reactions failed probably because of the close packing of the fibre chains and the strong hydrogen bonding between the peptide linkages not allowing accommodation for molecules of the size of vegetable tannins. However, by the introduction of bulky side chains or by imposing a certain irregularity in the composition of the amide chains a sufficient degree of reactivity towards tannins is obtained. Gustavson[24] thus found that a hydrated polyamide (a copolymer composed of 60% of adipic acid—hexamethylenediamine salt with 40% caprolactam) possessed a marked affinity for vegetable tannins and bound them irreversibly. Since the only reactive group of any importance in this polymer is the amide group, Gustavson[24] concluded that the peptide groups in collagen must similarly function as the "active sites" for binding of the vegetable tannins. Gustavson, in agreement with White, also suggested that the abundance of phenolic groups on the tannin molecules provided points of attachment and the steric possibilities for a multipoint combination with the peptide groups of adjacent protein chains.

IV. THE SCOPE OF VEGETABLE TANNIN CHEMISTRY

In spite of certain obvious inadequacies this qualitative description of the chemistry of leather manufacture none the less permits a fairly accurate definition of the principal objects of study in any examination of the vegetable tannins. The prime substances of interest are fairly obviously the polyphenolic compounds present in the plant extracts with the ability to convert hides and skins to leather, and if the views of White[21] and Gustavson[24] are accepted these are molecules with molecular weights in the range 500–3000. The chemical structure, physicochemical properties and biogenesis of these compounds are topics of undoubted importance and worthy of investigation. Vegetable tannin extracts also always contain many simpler polyphenols (such as the various monomeric flavonoid compounds and derivatives of the hydroxycinnamic acids) which impart a distinctive character to the tanning liquors, but which are ruled out from consideration as tannins by White's definition. It is not obvious, however, from previous work, how the presence of these minor components may affect the mechanism of the tannage by the tannins themselves and many of the earlier physicochemical studies appear to ignore their presence altogether. It seems probable that gallic acid (3), *m*-digallic acid (4) and the widely distributed chlorogenic acid (5), for instance, would readily penetrate the interfibrillar regions during tannage and compete with the vegetable tannins for active sites on the collagen molecule. In this respect,

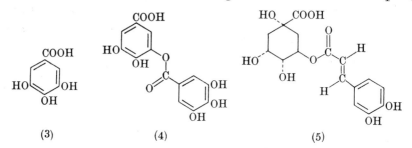

(3)   (4)   (5)

therefore, a detailed and systematic investigation of vegetable tannins should also include studies of these low molecular weight polyphenols, in particular their chemical structure, their effect on vegetable tannage, and their biosynthesis. In the case of the condensed tannins studies of this type are particularly relevant, since the monomolecular flavan-3,4-diols and flavan-3-ols, which frequently accompany these tannins in woody tissues, are considered[26, 27] to be their most likely chemical or biochemical precursors and here an understanding of the chemistry of these simpler phenols is implicit in an understanding of the tannins themselves.

V. CLASSIFICATION OF VEGETABLE TANNINS

The most acceptable division of the vegetable tannins is into the *hydrolysable* and *condensed* as originally suggested by Freudenberg[27] and is a classification based on structural types. The main distinctions between the two groups arise from their action towards hydrolytic agents, particularly acids. The hydrolysable tannins which have a poly-ester structure are readily hydrolysed by acids (or enzymes) into a sugar or related polyhydric alcohol and a phenol carboxylic acid and dependent on the nature of the latter a subdivision into *gallotannins* and *ellagitannins* is also usually made. Thus on hydrolysis the *gallo-tannins* give gallic acid (3) and the *ellagitannins* hexahydroxydiphenic acid (6) (isolated normally as its stable dilactone ellagic acid (7)), or acids which can be considered to be derived by simple chemical trans-formations of (6) such as oxidation, reduction and ring fission.

Schmidt's[28] elegant schemes of biosynthesis of the ellagitannins link the two groups since he considers the hexahydroxydiphenoyl group (8) characteristic of the ellagitannins to be derived by oxidative coupling of two suitably disposed galloyl residues in a gallotannin (Fig. 4). The

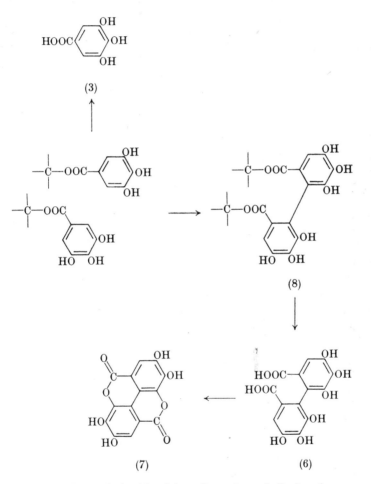

FIG. 4. Inter-relationship of the gallotannins and ellagitannins.

alcoholic portion of the hydrolysable tannins has most often been found to be D-glucose, although there also exist several authenticated cases of other sugars and in one case quinic acid acting as the core[29].

The condensed tannins in contrast do not readily break down with acid; instead they undergo progressive polymerization under the action of acids to yield the amorphous phlobaphens or tannin reds. The derivation of the tannins themselves is a matter of some conjecture although most workers agree that they are formed by a similar self condensation or polymerization of monomeric flavan-3,4-diol or flavan-3-ol precursors. Freudenberg[30] has suggested that their formation is a purely post-mortem process occurring over a long period of time and if this view is held then the condensed tannins, unlike the gallotannins and ellagitannins, are not direct products of plant metabolism. Other workers, whilst agreeing that further tannin polymerization may occur as a post-mortem process in the heartwood and bark, favour the view that the tannins are formed in living tissues under enzymic control and the consensus of opinion and such evidence as there is probably supports this idea.

REFERENCES

1. Kendrew, J. C., "The Proteins", p. 909 (Neurath, H. and Bailey, K., eds.), Academic Press, New York (1954).
2. Bear, R. S., *Adv. Protein Chem.*, **7**, 69 (1952).
3. Rich, A., and Crick, F. H. C., *J. molec. Biol.*, **3**, 483 (1961).
4. Ramachandran, G. N., and Kartha, G., *Nature, Lond.*, **176**, 593 (1955).
5. Harrington, W. F., and von Hippel, P. H., *Adv. Protein Chem.*, **16**, 69 (1961).
6. Astbury, W. T., *J. int. Soc. Leath. Trades Chem.*, **24**, 69 (1940).
7. Schroeder, W. A., Kay, L. M., Legette, J., Honnen, L., and Green, F. C., *J. Am. chem. Soc.*, **76**, 3556 (1954).
8. Kroner, T. D., Tabroff, W., and McGarr, J. J., *J. Am. chem. Soc.*, **77**, 3356 (1955).
9. Rich, A., and Green, D. W., *A. Rev. Biochem.*, **30**, 98 (1961).
10. Ramachandran, G. N., and Sasisekharan, V., *Nature, Lond.*, **190**, 1004 (1961).
11. Grassmann, W., Hannig, K., Endres, H., and Riedel, A., *Hoppe-Seyler's Z. physiol. Chem.*, **306**, 123 (1956).
12. Grassmann, W., Hannig, K., and Schleyer, A., *Hoppe-Seyler's Z. physiol. Chem.*, **322**, 71 (1960).
13. Schmitt, F. O., *Rev. mod. Phys.*, **31**, 349 (1959).
14. Hodge, A. J., and Schmitt, F. O., *Proc. natn. Acad. Sci. U.S.A.*, **46**, 186 (1960).
15. Schmitt, F. O., *Proc. Am. phil. Soc.*, **100**, 476 (1956).
16. Nishigai, M., Yagai, Y., and Noda, H., *J. Biochem., Tokyo*, **48**, 152 (1960).
17. Boedeker, H., and Doty, P., *J. Am. chem. Soc.*, **780**, 4267 (1956).
18. Hall, C. E., and Doty, P., *J. Am. chem. Soc.*, **80**, 1269 (1958).
19. Phillips, H., *Jl. R. Soc. Arts*, 102, 824 (1954).
20. Gustavson, K. H., "The Chemistry of Tanning Processes", Academic Press, New York (1956).
21. White, T., "The Chemistry of Vegetable Tannins", Soc. Leather Trades Chemists, Croydon, 1 (1956).
22. Grassmann, W., *Collegium, Haltingen*, **809**, 530 (1937).
23. Gustavson, K. H., and Holm, B., *J. Am. Leath. Chem. Ass.*, **47**, 700 (1952).

24. Gustavson, K. H., *J. Polym. Sci.*, **12**, 317 (1954).
25. Batzer, G., and Weissenberger, G., *Makromolek. Chem.*, **17**, 320 (1952).
26. Bate-Smith, E. C., and Swain, T., *Chemy Ind.*, 377 (1953).
27. Freudenberg, K., "Die Chemie der Naturlichen Gerbstoffe", Springer Verlag, Berlin (1920).
28. Schmidt, O. Th., and Mayer, W., *Angew. Chem.*, **68**, 103 (1956).
29. Haworth, R. D., and Haslam, E., "Progress in Organic Chemistry", Vol. 6 (Carruthers, W., and Cook, J. W., eds.), Butterworths, London (1964).
30. Freudenberg, K., *Experentia*, **16**, 101 (1960).

CHAPTER 2

Condensed Tannins—Analysis of the Extracts

I. Introduction

The traditionally important commercial sources of tannins are the heartwoods of *Schinopsis lorentzii* and *S. balansae* (Quebracho), the bark and heartwood of *Acacia mollisima* (syn. *A. mearnsii*) (Wattle or Mimosa) and the bark of *Rhizophora* species (Mangrove) but many dicotyledonous trees notably of the families *Leguminosae* (*Acacia* sp.), *Anacardiaceae* (*Schinopsis* sp.), *Myrtaceae* (*Eucalyptus* sp.) and *Fagaceae* (*Castanea* and *Quercus* sp.) are also remarkable for the large amounts of condensed tannins they contain (Table I).

TABLE I

SOURCES OF CONDENSED TANNINS[1]

Family	*Plant material*	*Yield* (%)
Myrtaceae	*Eucalyptus astringens* (bark)	40–50
	Eucalyptus wandoo (bark and heartwood)	12–15
Leguminosae	*Acacia catechu* (heartwood)	15
	Acacia mollisima (bark)	35–40
	Robinia pseudacacia (bark)	7
Anacardiaceae	*Schinopsis balansae* (heartwood)	20–25
	Schinopsis lorentzii (heartwood)	16–17
Rhizophoraceae	*Rhizophora candelaria* (bark)	25–30
	Rhizophora mangle (bark)	20–30
Fagaceae	*Castanea sativa* (bark)	8–14
	Quercus robur (bark)	12–16
Pinaceae	*Picea abies* (bark)	5–20
	Pinus sylvestris (bark)	16
	Larix decidua (bark)	5–20

Routine estimations of tannin concentrations in commercial extracts such as those shown in Table I and also Chapter 4, Table I, are normally performed by a weight-difference method[2]. An aqueous infusion is prepared and the total solids determined by evaporation of an aliquot. The procedure is then repeated with a further aliquot after removal of the tannins by adsorption onto hide or polyamide powder and comparison of the two figures allows the tannin to non-tannin ratio in the extract to be determined. Grassmann and Zeschitz[3] have described an alternative micro-method for the estimation of tannins in which the latter are preferentially adsorbed from aqueous solution onto a cellophane membrane and then determined colorimetrically by complex formation with methylene-blue.

The chemical literature concerning the structure and formation of the condensed tannins contains many speculative theories which have been adequately summarized and criticized by other workers[4]. A weight of evidence has, however, accumulated in recent years which favours the view, originally suggested in the form of the "catechin hypothesis" by Freudenberg,[5] that the complex polymeric structures of many of the condensed tannins are derived primarily by polycondensation of precursors of a flavonoid type. The generic term catechin was first used by Freudenberg to describe the colourless crystalline substances commonly located in plant tissues in association with the condensed tannins. Since the catechins were readily converted *in vitro* to amorphous tannin-like materials Freudenberg therefore regarded them as direct precursors of the tannins in Nature. Later work has shown the catechins to be derivatives of the basic flavan-3-ol structure and nowadays these compounds, to avoid confusion with trivial names, are generally referred to as flavan-3-ols. The first and perhaps most widely distributed compounds of this class to be isolated from natural sources[6, 7] were (+)-catechin from *Uncaria gambir* and (−)-epicatechin from *Acacia catechu* and Freudenberg showed[8, 9] these to be diastereoisomers of the 5,7,3',4'-tetrahydroxyflavan-3-ol structure (1 ; R = OH). More recent work has emphasized the generality of this simultaneous natural occurrence of condensed tannins with members of the flavan-3-ol group of compounds and has led also to the additional recognition[10] of flavan-3,4-diols (such as 2 ; R = OH) as possible precursors of these tannins in Nature.

The production of red colours from colourless substances which are present in the leaves, flowers and fruit of certain plants upon treatment with hot acid has long been known ; Robert Boyle being accredited with one of the first observations over a hundred years ago[11]. Rosenheim[12] during a study of the anthocyanin pigments of the leaves of the young grape vine was probably the first to recognize a compound of this type which, since it gave cyanidin (3; R = H) with acid, he named leucocyanin. Its structure, in view of this characteristic reaction, he suggested was that of a glycoside of the pseudo-base of cyanidin (4; R = sugar). The widespread natural distribution of these compounds was, however, not revealed until the extensive surveys of plant species were made by Robinson and Robinson[13] and later Bate-Smith[14, 15]. Both groups of workers drew attention to the ubiquity in plants of leucoanthocyanins especially those which gave cyanidin (3; R = H) and delphinidin (3; R = OH) upon acid treatment and Bate-Smith observed that in general their occurrence was more prevalent in plant species with a woody habit of growth. In the light of subsequent structural studies many of these leucoanthocyanins (or in some texts leucoanthocyanidins) are more systematically described as flavan-3,4-diols (such as 2; R = OH) and this terminology is used wherever it is applicable in the following discussion.

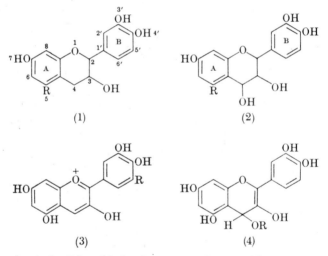

(1)   (2)   (3)   (4)

When heated with acid the flavan-3,4-diols yield varying amounts of amorphous polymers in addition to the corresponding anthocyanidin and in this respect their behaviour closely resembles that of the condensed tannins themselves which under similar conditions give amorphous "tannin reds" of phlobaphens. This characteristic reaction of flavan-3,4-diols and their frequent association in plant tissues with condensed tannins prompted Bate-Smith and Swain[10] to suggest that flavan-3,4-diols as well as flavan-3-ols acted in Nature as precursors of these tannins. The subsequent observations of workers such as Roux[16] and his collaborators has borne out the fruitful nature of this suggestion and has fully illustrated the central role of flavan-3,4-diols in the production of condensed tannins in plant tissues.

It is probable following the work of Hathway[17], Grassmann and his colleagues[18] and White and King[19] that other simple polyphenols besides the flavan-3-ols and flavan-3,4-diols participate in condensed tannin formation, although in general the evidence is not so formidable or convincing as for the hydroxyflavan derivatives. Grassmann has put forward the suggestion that spruce-bark tannin is derived primarily by polymerization of piceatannol (revised structure[20] 5; R = OH) which either free or as its mono- or diglucoside is the major polyphenol in the inner bark. Hathway and Seakins[17] have similarly isolated resveratrol (5; R = H) and its 3-β-D-glucoside from the tanning bearing heartwood of *Eucalyptus wandoo* and on the basis of Grassmann's ideas these stilbenes may also be expected to be implicated in condensed tannin formation in this wood. White and King[19] have put forward a more

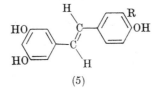

(5)

radical view concerning the formation of condensed tannins in plants from a study of phenolic metabolism in *Schinopsis* species. The evidence they suggested showed that the hydrolysable tannins were the primary phenolic products in the leaves and that these compounds were then translocated to the sapwood–heartwood boundary where they took part in condensed tannin formation. Despite the fact that recent evidence[21] supports the idea that polyphenols found in the wood are not translocated but are formed *in situ* from carbohydrates these ideas and those of Grassmann are worthy of further and more thorough consideration since they emphasize the view that the structural patterns in condensed tannins from different plant species may not necessarily be the same.

II. PREPARATION OF PLANT EXTRACTS

Condensed tannins are located in quantity primarily[1] in the bark and heartwood of dicotyledonous plants and whatever method of extraction is used the final extract generally contains a range of low molecular weight phenolic components in addition to the tannins themselves. Fractionation may then be undertaken with one or more aims in view, either to separate each individual compound for further structural study or merely to isolate the tannins. In the case of the condensed tannins efforts which fall within the latter category have so far met with little success and in consequence actual degradative work on the tannins has been limited. In contrast the isolation and identification of the simpler phenols in these extracts has been a more profitable line of study and has led to useful working hypotheses regarding the structure of the tannins and to the discovery of many interesting biogenetic inter-relationships. The consensus of information suggests that the condensed tannins are derived by a *simulated* acid catalysed condensation or oxidative polymerization of flavan-3-ol or flavan-3,4-diol precursors and attention is directed in particular here to the isolation and structural analysis of these substances. It is also considered useful, in view of the scanty knowledge of the actual structural patterns possessed by the tannins themselves to examine the possibility that other monomeric phenols may be involved in their formation and hence methods for the identification of some additional minor components of tannin extracts—in particular other flavonoids and stilbenes—are also briefly discussed. For the organic chemist this wider approach has the further advantage that it permits a deeper insight into the biogenetic relationships between the various phenolic compounds in the plant tissue.

In the tannery the tanner acquires his raw material in the natural state from which the tannins are leached as required, or as a liquid or solid concentrate produced by prior hot water extraction of the tannin bearing material. In view of the known reactivity of some of the simpler phenolic constituents these latter conditions must be regarded as leading inevitably to a greater complexity in the extract. For the studies to have any biogenetic significance it is necessary for the extract to contain the tannins and other substances approximating closely in chemical constitution to those in the plant tissues and wherever possible therefore work should begin with the fresh plant itself, or a part of it. Generally acetone, methanol or ethyl acetate extracts prepared by *cold* solvent percolation of crushed plant material or by rapid extraction in a high-speed mixer give for this purpose the most satisfactory results.

III. QUALITATIVE ANALYSIS OF FLAVONOID COMPOUNDS

Paper chromatography has proved to be a powerful analytical technique when used in the study of plant extracts[22] and has been applied extensively in the preliminary qualitative analysis of condensed tannin materials. Although the tannins themselves give poorly defined patterns

(Fig. 1) paper chromatographic methods are ideal for the analysis of the co-occurring flavonoid compounds since these have a wide range of solubilities and are readily detected on chromatograms by means of specific spray reagents and in certain cases by their distinctive fluorescences in ultra-violet light (Table II). Numerous solvent systems have been employed in the investigation, paper chromatographically, of plant extracts but the usual pair employed in a two-dimensional examination consists of dilute acetic acid for the first direction of development and a partitioning mixture—normally a combination of butan-2-ol,* acetic acid and water—for the second. The use of solvent pairs

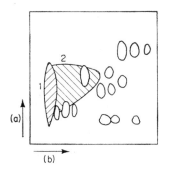

FIG. 1. Paper chromatogram of quebracho extract. Solvents (a) 6% acetic acid; (b) butan-2-ol–acetic acid–water (14:1:5). "Tannins"—areas 1 and 2.

of this type offers the advantage of a separation based on different physical principles for each direction of development of the chromatogram. In dilute acetic acid separations are ascribed both to differential adsorption by the cellulose and to differences in solubility in the aqueous medium. In the predominantly organic phase separations are due to the repeated partitioning of the solute between the solvent and the microscopic stationary aqueous phase of the paper. The wealth of experimental data[22, 23] on the paper chromatography of flavonoid compounds which is available has inspired several attempts[22, 24] to assess the structural features which are operative in determining R_f values (Table III). The significant immobility in the aqueous phase of development of the aglycones of flavones, flavonols, anthocyanidins, aurones and chalcones has been interpreted both in terms of the low solubility of these compounds in water and to their planar configuration which permits adsorption (via hydroxyl groups) to the cellulose support at

* Butan-2-ol is preferable to butan-1-ol since it undergoes less ready esterification and the composition of the solvent phase remains constant over long periods.

TABLE II

COLOUR REACTIONS OF FLAVONOID COMPOUNDS

Reagent	Light	Flavone	Flavonol	Isoflavone	Flavanone	Aurone	Chalcone
None	visible	pale yellow	yellow	—	—	brown-yellow	yellow
None	u.v.	dull brown	yellow	weak purple	—	brown-yellow	dull brown
NH_3	u.v.	green	yellow	weak purple	pale yellow	brown-red	dull red
$AlCl_3$	visible	pale yellow	yellow	—	—	pale yellow	yellow
$AlCl_3$	u.v.	green*	yellow*	yellow*	green*	green*	orange*
$NaBH_4$	visible	—	—	—	magenta	—	—

* Fluorescent.

several points simultaneously. Directly the planar character of the flavonoid molecule is disturbed by attachment of sugar residues or reduction of the heterocyclic ring system (cf. flavan-3,4-diols, flavan-3-ols and 2,3-dihydroflavonols) mobility in the aqueous phase results and this is probably due to a certain increased water solubility and also to a decreased affinity for the cellulose. In the organic partitioning phase the R_f values within a particular class of flavonoids decreases with an increase in the content of phenolic or alcoholic hydroxyl groups. In the aqueous phase of development phenolic and alcoholic hydroxyl groups have opposing effects on the R_f value of a flavonoid molecule; in general increase in the former lowers the R_f but increase in the content of alcoholic hydroxyl groups has the reverse effect. Examples of these trends and the effects of structure on R_f are seen by reference to Table III.

The reduced state of the heterocyclic ring system in the 2,3-dihydro-flavanols, flavan-3-ols and flavan-3,4-diols permits in these compounds different spatial arrangements of the groups attached at the 2, 3 and 4 positions. The absolute configuration of many of these compounds has been rigidly proved and is conveniently expressed in terms of the nomenclature of Cahn, Ingold and Prelog[25]. Thus (+)-catechin (6; R = H) and (−)-epicatechin (7; R = H) are represented[26] as the *trans* (2R, 3S) and *cis* (2R, 3R) configurations respectively; (+)-gallo-catechin and (−)-epigallocatechin are analogously written as (6; R = OH) and (7; R = OH). Roberts and Wood[27] were the first to show that diastereoisomers such as (+)-catechin and (−)-epicatechin which differ only in configuration at C-3 could be separated by paper

TABLE III

R_f VALUES OF FLAVONOID COMPOUNDS

Class	Compound	Pattern of hydroxylation	Colour with bis-diazotized benzidine	B.A.W. R_f	2% Acetic acid R_f
Flavonol	Kampferol	3,5,7,4'	—	0·83[a]	0·0[a]
	Kampferol-3-glucoside	5,7,4'	—	0·70[a]	0·43[a]
	Fisetin	3,7,3',4'	pale yellow	0·73[a]	0·0
	Quercetin	3,5,7,3',4'	red-yellow	0·64[a]	0·0
	Quercetin-3-glucoside	5,7,3',4'	red-yellow	0·58[a]	0·37[d]
	Robinetin	3,7,3',4',5'	golden brown	0·40[a]	0·0
	Myricetin	3,5,7,3',4',5'	golden brown	0·43[a]	0·0
2,3-Dihydroflavonol	(+)-Fustin	3,7,3',4'	pale yellow	0·65[c]	0·37
	(−)-Fustin	3,7,3',4'	pale yellow	—	0·35
	(+)-Taxifolin	3,5,7,3',4'	red-yellow	0·78[b]	0·28
	(+)-Dihydrorobinetin	3,7,3',4',5'	golden yellow	0·50[c]	0·34
Flavan-3-ol	(−)-Fisetinidol	3,7,3',4'	pale yellow	—	0·48
	(+)-Fisetinidol	3,7,3',4'	pale yellow	—	0·43
	(+)-Robinetinidol	3,7,3',4',5'	canary yellow	0·50[c]	0·42
	(+)-Catechin	3,5,7,3',4'	claret-maroon	0·75[b]	0·35
	(−)-Epicatechin	3,5,7,3',4'	claret-maroon	0·65[b]	0·30
	(−)-Epicatechin-3-gallate	5,7,3',4',3",4",5"	claret-maroon	0·86[b]	0·23
	(+)-Gallocatechin	3,5,7,3',4',5'	claret-maroon	0·57[b]	0·32
	(−)-Epigallocatechin	3,5,7,3',4',5'	claret-maroon	0·47[b]	0·24
	(−)-Epigallocatechin-3-gallate	5,7,3',4',5',3",4",5"	claret-maroon	0·72[b]	0·20
Flavan-3,4-diol	(+)-Leucofisetinidin	3,4,7,3',4'	pale yellow	—	0·52
	(−)-Leucofisetinidin	3,4,7,3',4'	pale yellow	—	0·47
	(+)-Leucorobinetinidin	3,4,7,3',4',5'	canary yellow	0·59[c]	0·46
	(−)-Leucorobinetinidin	3,4,7,3',4',5'	canary yellow	—	0·40

B.A.W.—Butan-1-ol–acetic acid–water. [a] 4:1:5; [b] 4:1:2·2; [c] Butan-1-ol saturated with water; [d] 15% aqueous acetic acid.

chromatography using water as irrigant, and several theories have been
advanced to account for this phenomenon.

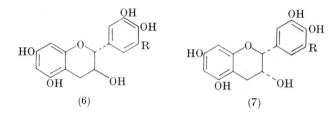

On the basis of their infra-red spectra, which show the 3-hydroxyl
group to be hydrogen bonded (νOH $-$ 3590 cm^{-1}), ($+$)-catechin and
($-$)-epicatechin tetramethyl ethers have been assigned[26] the conformations (8; R $=$ Me) and (9; R $=$ Me) respectively, with the C-3 hydroxyl
group placed in an axial position in the half chair conformation of the
heterocyclic ring and hydrogen bonded to the heterocyclic oxygen. In
support of these assignments Mehta and Whalley[28] have pointed out
that the 1,2-shifts so characteristic of catechin and its derivatives
(see p. 43) require the 2-aryl and 3-hydroxyl groups to be in the *trans*
diaxial relationship (8) in the transition state and that this conformation may well be preferred in the ground state of these compounds.
Philbin and Wheeler[29] have alternatively interpreted the infra-red
spectra of some flavan-3,4-diols in terms of conformations in which the
C-3 hydroxyl group is engaged in OH$\cdots\pi$ bonding to the aromatic
ring B and if these arguments are extended to ($+$)-catechin and
($-$)-epicatechin methyl ethers the respective conformations (10,
R $=$ Me and 11, R $=$ Me) result. In these the dihydropyran ring is
illustrated in the "sofa" conformation, a modification of the half chair
suggested by Philbin and Wheeler[30] where atoms 1, 3 and 4 are co-
planar with ring A. Adoption of the latter type of hydrogen bonding
allows the formulation of the ($+$)-catechin derivative with the 2-aryl
group (ring B) in the conformationally more favourable equatorial
position (10). For effective hydrogen bonding in ($-$)-epicatechin tetramethyl ether rings A and B are approximately co-planar (11), but for
($+$)-catechin tetramethyl ether rings A and B must lie in mutually
orthogonal planes. Whichever interpretation of the infra-red and
chemical data is favoured both predict a more compact molecular
structure for ($-$)-epicatechin (9 or 11) than for ($+$)-catechin deriva
tives (8 or 10). The higher R_f values of ($+$)-catechin compared to
($-$)-epicatechin may then be interpreted in terms of the overall more
planar structure of ($-$)-epicatechin (9 or 11) lending itself to stronger
adsorption by the cellulose. The value of this approach is of course

limited since the infra-red spectra were measured in aprotic solvents and a quite different state of affairs may well prevail in the hydroxylic media used for chromatography. In these circumstances a simpler explanation of these characteristic differences in R_f values is that in (+)-catechin (10) the equatorial C-3 hydroxyl group is much more readily solvated than the corresponding axially orientated hydroxyl group in (−)-epicatechin (11). Roberts and Wood[27] also demonstrated in their work with the flavan-3-ols of tea that (±)-catechin and (±)-gallocatechin were resolvable into their optical antipodes by paper

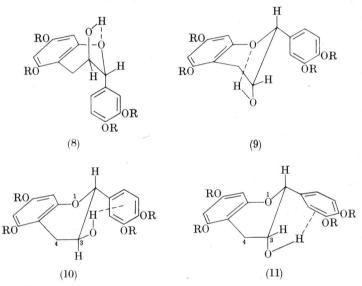

limited since the infra-red spectra were measured in aprotic solvents
chromatography using water as solvent and presumably in these separations the cellulose acts as an asymmetric adsorbent. Resolutions such as these and the separations of the diastereoisomeric flavan-3-ols illustrate the elegance of the paper chromatographic method of analysis when applied to pure compounds. However, it is dangerous to assume that separations of this order are therefore possible in the analysis of complex plant extracts since more often than not the R_f values of individual substances are influenced considerably by the presence of other compounds. Under these conditions small differences in R_f such as those between (+)- and (−)-catechin in water are too small to allow more than tentative identifications by paper chromatographic methods alone.

Detection of flavonoid compounds on paper chromatograms may be achieved by numerous methods. In common with other polyphenols many absorb or fluoresce in ultra-violet light[22], particularly those with a carbonyl group at position 4 on the heterocyclic ring. This fluorescence is frequently enhanced by formation of the corresponding phenolate anion (fuming with ammonia vapour) or by chelation (spraying with ethanolic aluminium chloride or methanolic citric–boric acid solutions); examples of these effects are shown in Table II. Various sprays [22, 23, 31] are also available for the identification of typical orientations of hydroxyl groupings in flavonoids and a selection of these are discussed briefly below.

1. FERRIC CHLORIDE—POTASSIUM FERRICYANIDE[32]

A freshly prepared spray consisting of equal parts of 2% ferric chloride and 2% potassium ferricyanide containing a drop of potassium permangate solution is used. Excess reagent is removed by washing the chromatogram in dilute hydrochloric acid and then water, and the chromatogram dried in air. o- and p-Dihydroxy, o-trihydroxy and other easily oxidized phenols such as arbutin (hydroquinone-β-D-glucoside) react to give a prussian blue spot on a white background.

2. AMMONIACAL SILVER NITRATE[33]

The spray is prepared by addition of 6 N ammonia to silver nitrate solution (14 g in 100 ml water) until the silver oxide formed just redissolves. The chromatogram is washed, after spraying, with sodium thiosulphate solution (0·1%) and water, then air dried. Readily oxidized phenols as above are recorded as brown-black spots on a white background.

3. BIS-DIAZOTIZED BENZIDINE[34]

The spray, which is used immediately after preparation, contains two parts of a solution of benzidine hydrochloride (6 g in concentrated hydrochloric acid (14 ml) and water (980 ml)) and three parts of a 10% sodium nitrite solution. After allowing time for reaction (3 min) chromatograms are washed in water. o-Dihydroxy phenols are shown as yellow spots, m-dihydroxy phenols as claret-maroon areas on the chromatogram (Table III). Benzidine has achieved some notoriety as a carcinogen and diazotized p-nitroaniline may be usefully used in its place.

4. p-TOLUENESULPHONIC ACID[35]

Chromatograms are lightly sprayed with a 3% solution of p-toluene-sulphonic acid in ethanol and then heated at 80–100°C for 5 minutes. Flavan-3,4-diols are detected as rose-pink spots, flavan-3-ols as buff to brown spots.

Sprays (1) and (2) may be used primarily to investigate the number and orientation of hydroxyl groups in ring B of the flavonoid. The bis-diazotized benzidine spray (3) combines reactivity towards both m- and o-dihydroxyphenols and where both groupings are present in the same molecule its reaction with the m-dihydroxy group (generally ring A) is most significant in producing the final colour. This effect is most pronounced in flavonoids where there is no carbonyl group at position 4 and hence when ring A is not deactivated towards electro-philic substitution (e.g. flavan-3-ols, flavan-3,4-diols, Table III). The presence of flavan-3,4-diols (spray 4) in an extract may be corroborated by the use of the original method of Bate-Smith[14] for the detection of these compounds. A sample of the extract is heated with 2 N hydro-chloric acid at 100°C for 20 minutes when the flavan-3,4-diol is partially converted to the corresponding anthocyanidin which may be detected (Table IV) spectrophotometrically or by paper chromatography. Bate-Smith's method for the production of the anthocyanidin is less suitable for condensed tannin extracts than for plant tissues to which it was originally applied. With condensed tannin extracts subsequent paper chromatographic identification of the anthocyanidin is often made difficult due to the simultaneous production of phlobaphens from the tannin when it is treated with acid. An improved procedure which restricts phlobaphen formation is that of Pigman and co-workers[36] in which the extract is heated for a short period at 100°C under pressure (2–3 atm) with 0·6 N hydrochloric acid in propan-2-ol or butan-1-ol.

Roux and Maihs[31] have extended these paper chromatographic

TABLE IV

PAPER CHROMATOGRAPHIC AND SPECTROPHOTOMETRIC IDENTIFICATION OF ANTHOCYANIDINS[37]

Anthocyanidin	Pattern of hydroxylation	λ_{max} mμ MeOH/HCl	R_f values Forestal	B.A.W.
Apigenidin	5,7,4'	476	0·75	0·74
Luteolinidin	5,7,3',4'	493	0·61	0·56
Pelargonidin	3,5,7,4'	520	0·68	0·80
Cyanidin	3,5,7,3',4'	535	0·50	0·68
Paeonidin	3,5,7,4',3'-OMe	532	0·63	0·71
Delphinidin	3,5,7,3',4',5'	546	0·30	0·42
Petunidin	3,5,7,4',5',3'-OMe	546	0·46	0·52

Forestal—Water–acetic acid–concentrated hydrochloric acid, 10 : 30 : 3.
B.A.W.—Butan-1-ol–acetic acid–water, 4 : 1 : 5.

methods to the semi-quantitative estimation of flavonoid compounds in tannin extracts. Chromatograms when developed are sprayed (2 or 3 above) and the colours produced compared with standards by means of a densitometer. Using this technique Roux has estimated the variation in concentration of flavonoid compounds in different tissues of the same plant and in relation to the biogenesis of the condensed tannins he has made the important observation that the concentration of flavan-3,4-diols decreases with increasing age of the heartwood of a tree.

Paper chromatographic methods provide perhaps the most suitable techniques for the analysis of plant polyphenols with a predominantly hydrophilic character. In the case of simpler phenols which have a significant lyophilic nature other chromatographic methods of identification may be utilized with some advantage. Thin layer chromatography has many points of superiority, such as speed, sensitivity and efficiency of separation when compared to paper chromatography for the analysis of many types of compound and its development in the past few years has provided the chemist with important new methods of analysis[38]. The method has not been extensively used as yet for the identification of phenolic compounds in plant extracts but a variety of simple phenols—hydroxybenzoic and cinnamic acids[39, 40], coumarins[41], anthocyanidins[40, 42], flavones[41] and the simpler glycosides such as arbutin have been separated on thin layers of silica gel, kieselguhr and polyamide. The methods used in the preparation of the chromatoplates and for the detection of phenols have been reviewed by Randerath[38]. The rate of migration of a phenolic compound on silica gel is determined primarily by the number and position of hydroxyl groups in the molecule (Table V) and similar relationships hold for chromatography of phenols on polyamide and kieselguhr. It is interesting to note that chromatography of phenols on polyamide powder is dependent on the formation of hydrogen bonds between the phenolic hydroxyl groups and the amide groups of the polymer (simulating to some extent the natural process of tannage), and the ability of a particular solvent to displace the phenols from their points of attachment to the amide linkage. It seems highly probable that the methods of thin-layer chromatography will be used to a greater extent in future work for the analysis of plant phenols in some cases to supplement and in others to replace existing techniques.

Other techniques which have been used for the analysis of plant phenols are those of gas liquid chromatography[43] and paper electrophoresis[44], but both of these are of limited applicability when compared to paper and thin layer chromatography. Narasimachari and von Rudloff[43] demonstrated that some polyphenols including flavonoids

TABLE V

THIN LAYER CHROMATOGRAPHY OF PHENOLS ON SILICA GEL[39]

Compound	R_f*
Phenol	0·60
Catechol	0·54
Protocatechuic acid	0·39
Gallic acid	0·23
Syringic acid	0·60
p-Coumaric acid	0·52
Caffeic acid	0·43
Ferulic acid	0·58

* Benzene–methanol–acetic acid, 45 : 8 : 4.

may be successfully gas chromatographed using a silicone polymer as the liquid phase. The technique they suggested might also be used for the isolation and identification of these compounds but in their work they were not able to establish a clear relationship between the retention times and the number and position of substituents on the flavonoid nucleus. In general the gas chromatographic separations of these polyphenols had to be carried out at fairly high temperatures (235°C) to ensure reasonable minimum retention times and this resulted in several cases in the promotion of chemical reactions on the column. Thus flavanones and chalcones gave two distinct peaks on the gas chromatogram which were attributed to isomerization of these compounds during the analysis. In an extension of this work Sato and von Rudloff[45] showed that free phenolic compounds may also be readily separated by gas chromatography as their trimethylsilyl ethers which in contrast to the free phenols are relatively volatile. The technique permits qualitative analysis of certain heartwood polyphenol fractions as Sato and von Rudloff[45] showed in an examination of the extract of red pine heartwood (*Pinus resinosa*). Paper electrophoretic measurements of plant polyphenols have not been widely used although again the method is a useful one for the analysis of the simpler phenolic compounds[44].

Combination of the observations made by the various methods of analysis described above often allows tentative identification of components of plant extracts to be made. However, it is desirable in so far as this is possible to follow up these preliminary analyses by the isolation and identification of compounds.

IV. ISOLATION OF FLAVONOID COMPOUNDS FROM PLANT EXTRACTS

Particular components of a plant extract which are present in a relatively high concentration may be isolated by direct fractional crystallization. Many examples of isolation by such methods are known and have been described[46, 47], thus acetone extraction of the heart-wood of *Acacia catechu* yields a gum from which (−)-epicatechin crystallizes[9] and (+)-catechin is also fairly readily obtained[48] from the ether extract of *Uncaria gambir*. Where, however, it is necessary to isolate and identify as many components as possible in a plant extract in order to elucidate structural, stereochemical and biogenetic relationships a more general approach is required.

Paper chromatographic analysis in addition to being an efficient method by which the complexity of the tannin extract may be assessed also provides a useful guide to the selection of a particular isolation procedure. Thus substances which are separable in the aqueous phase of paper chromatographic analysis can usually be separated by column or thin layer chromatography on cellulose. Mayer and Merger[49] utilized this principle in separating (+)- and (−)-catechin (R_f (H_2O); 0·35 and 0·29 respectively) and (+)- and (−)-epicatechin (R_f (H_2O); 0·25 and 0·30 respectively) from their racemates by chromatography in water on cellulose. Analogously a mixture of substances whose R_f values in the organic partitioning phase of paper chromatographic analysis are quite different may be separated by column or thin layer chromatography on cellulose using a suitable organic solvent as eluant, or, since paper chromatographic separations in the organic partitioning phase arise essentially from repeated partitions of the counter-current type, by counter-current distribution between water and an organic solvent such as ether, ethyl acetate or butan-1-ol. Where the R_f values of the substances in the aqueous phase of paper chromatographic development are approximately the same the distribution coefficients in the counter-current distribution increase with increasing value of the R_f in the organic phase of development. The isolation by Weinges[50] of (−)-epicatechin, (+)-catechin and (+)-taxifolin (R_f values, Table III) from an extract of the bark of *Pseudotsuga taxifolia* by counter-current distribution (500 transfers between ether and water) provides a good illustration of the principles of the method. Hörhammer and his colleagues[51] have similarly made extensive use of this technique in the separation and identification of other flavonols and their glycosides.

The inherent complexity of most condensed tannin extracts, however, makes it necessary to apply as a standard approach a combination of both chromatographic and counter-current distribution techniques for the isolation of its constituents. Roux and his colleagues have made considerable use of this synthesis of methods in detailed analyses of the flavonoid and tannin components of the woods of *Robinia pseudacacia*[52], *Schinopsis*[53] and *Acacia*[54, 55] species. Results of a typical extraction and fractionation of *Robinia pseudacacia*[52] are shown in Fig. 2 and Table VI. An initial separation of compounds in the crude methanolic extract (100 g) (paper chromatogram Fig. 2) was obtained by counter-current distribution (160 transfers) in the two phase system butan-2-ol–light petroleum–water (3 : 2 : 5). The tube contents were analysed by paper chromatography before grouping into eight major fractions (Table VI).

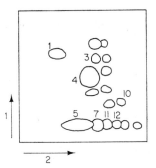

FIG. 2. Paper chromatogram of methanolic extract of heartwood of *Robinia pseud-acacia*[52]. Solvents (1) −2% acetic acid; (2) water saturated butan-1-ol. Compounds identified: 1, (+)-leucorobinetinidin; 3, (−)-robinetinidol; 4, (+)-dihydrobinetin; 5, robinetin; 7, robtein; 9, (+)-fustin; 10, (−)-butin; 11, fisetin; 12, butein.

Flavonoid compounds present in each of the fractions were then isolated by preparative thick paper chromatography (Whatman No. 3 paper with 2% acetic acid or butan-1-ol–acetic acid–water, 6 : 1 : 2, as developing solvent). [As a variant in future work thin layer chromatography on silica or polyamide might profitably be exploited at this stage of final purification as an alternative to paper chromatography.]

Other workers have employed chromatography on other adsorbents such as silica gel[56, 57], magnesol[58] or polyamide[59] to isolate flavonoid compounds from plant extracts. Thus Bradfield[56] and later Roberts[57] separated the various flavan-3-ols and their gallate esters from the other polyphenols of plucked tea shoots by chromatography on silica gel. Using similar methods Hathway[60] made an examination of oak bark tannins and Mayer[61] isolated (+)-catechin and (+)-gallocatechin from sweet chestnut extract. Hörhammer, Wagner and Leeb[59] have

TABLE VI

SEPARATION OF FLAVONOIDS IN METHANOL EXTRACT (100 g) FROM
HEARTWOOD OF *Robinia pseudacacia*[52]

Tube in countercurrent distribution	Compounds isolated
16–35	(+)-leucorobinetinidin (2·3 g) (flavan-3,4-diol)
36–64	
65–90	(+)-dihydrorobinetin (20 g) (2,3-dihydroflavonol) — (−)-robinetinidol (0·036 g) (flavan-3-ol)
91–107	
108–123	(±)-robtin (0·48 g) (flavanone) — (+)-fustin (0·07 g) (2,3-dihydroflavonol)
124–145	robinetin (1·4 g) (flavonol)
146–160	robtein (0·40 g); (±)-butin (0·165 g); butein (0·07 g) (chalcone) (flavanone) (chalcone) fisetin (0·01 g) (flavonol)

similarly used polyamide chromatography for the separation of flavones
and their glycosides and analogous separations were achieved by
Wender[58] using the magnesium silicate—magnesol. The relative
involatility of natural phenols does not make them particularly suitable
for separation by vapour phase chromatography but Narasimachari
and von Rudloff[43] utilized this technique in separating the hydroxy-
diphenyls aucuparin (12; R = H) and methoxyaucuparin (12; R =
OMe) from the wood of *Sorbus decora* and the method may be of use in
isolating the more volatile flavonoid components of other heartwoods.

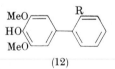

(12)

V. IDENTIFICATION OF FLAVONOID COMPOUNDS— PHYSICAL METHODS

Physical methods of structure determination play an increasingly
important role in the various fields of organic chemistry and this is

particularly evident in the flavonoid group of compounds. Owing to the difficulties of isolation, particularly of the minor components of tannin extracts, small quantities of such compounds are frequently all that are available for subsequent structural analysis. The characterization and identification of these substances has, however, been made considerably easier by the frequent application of physical methods of analysis[62] which in some cases permit unequivocal structural assignments or, more generally, direct the choice of definitive chemical degradations which need to be carried out to confirm one structure from a number of alternatives. In most cases an initial classification of a flavonoid compound is possible on the basis of paper or thin layer chromatographic behaviour[22], colour reactions[63], ultra-violet[64] and infra-red spectral

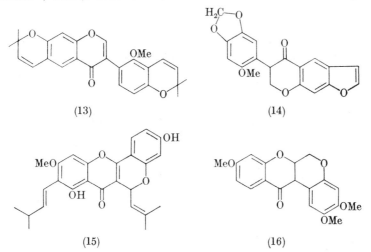

(13)   (14)

(15)   (16)

measurements[65] and more detailed structural analysis can then be made by use of the sophisticated proton magnetic resonance and mass spectral methods. Increasing use has been made particularly of proton magnetic resonance spectra in the determination of the structure of flavonoids and related compounds[66] and it has been applied effectively in solving the structures of some of the more complex members of this class such as munetone (13) (*Mundulea suberosa*)[67], neotenone (14) (bark, *Neorautenenia pseudopacchyrrhizus*)[68], cycloartocarpin (15) (heartwood, *Artocarpus integrifolia*)[69], munduserone (16) and sericetin (17a or b) (bark, *Mundulea sericea*)[70, 71], dalbergione (18) and related compounds (heartwood, *Dalbergia* species)[72] and ichthynone (19) (Jamaican dogwood, *Piscidia erythrina*)[73, 74]. The structures of many of these compounds contain a basic flavone or isoflavone structure which has undergone further elaboration by the addition in various forms of isoprenoid units and it is interesting to note in passing that as yet the analogous flavan-3-ols and flavan-3,4-diols have not been isolated from plant materials. This may be due to difficulties of isolation but their absence may have a biogenetic significance.

Proton magnetic resonance studies are also admirably suited to an investigation of the stereochemistry of the heterocyclic ring system of flavan-3,4-diols and several groups have reported on these features[75-78]. Crombie and Lown[79] have also made a detailed study of the proton magnetic resonance spectra of the rotenoids, e.g. rotenone (20) (root, *Derris elliptica*) and munduserone (16) in relation both to

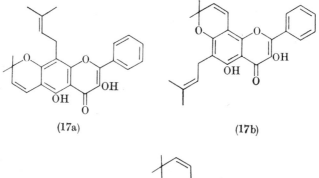

(17a)   (17b)

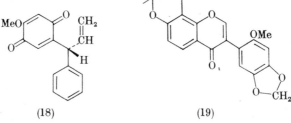

(18)   (19)

their structure and stereochemistry. They have shown that considerable structural information regarding for instance the substitution of rings A and D and the geometry of the B/C ring fusion is readily accessible by this means. A clear indication of the relative stereochemistry of the rotenoids is also provided by optical rotary dispersion measurements which combined with the results of chemical degradation allow the absolute configuration of these substances to be defined. The application of measurements of mass spectra although as yet not widely reported may be expected to be used to an increasing extent in the determination of flavonoid structures and the typical cracking patterns resulting from a number of simple flavones, isoflavones and rotenoids have been recorded by Reed and Wilson[80]. Thus for example apigenin

(21; R = H) and acacetin (21; R = Me) gave the parent molecular ion as the base peak and an abundant fragment ion corresponding to the loss of carbon monoxide. Fragmentations which were much less abundant arose by fission of the heterocyclic ring (22 and 23; R = H and Me). A more detailed discussion of some of these physical methods of structure determination as applied to the identification of flavonoids is given below.

Measurements of the infra-red spectra of flavonoid compounds have been widely used in interpreting structural features since a considerable amount of data is available for reference purposes[65, 81, 82]. The method

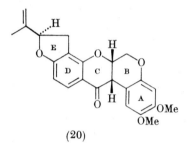

(20)

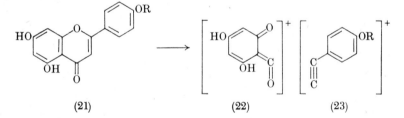

(21)   (22)   (23)

generally provides useful information regarding the state of oxidation of the heterocyclic ring system and also of the presence of isoprenoid units attached to the flavone nucleus. The presence of carbonyl absorption is, apart from the gallate esters of (−)-epicatechin and (−)-epigallocatechin, diagnostic for flavones, isoflavones, 2,3-dihydroflavones, chalcones and aurones. In general the carbonyl stretching frequency is higher (1650–1700 cm^{-1}) for 2,3-dihydroflavones than for flavones and isoflavones where the planar nature of the heterocyclic ring system permits increased contributions to the resonance hybrid of structures such as (24). The carbonyl absorption of aurones (25) also occurs at higher frequencies than that of flavones and in this case this is due to the increased angular strain in the five membered ring system. Substituents in certain positions of the flavone nucleus have a marked effect on the carbonyl absorption (Table VII). Thus a hydroxyl in the 3

2*

TABLE VII

CARBONYL STRETCHING FREQUENCIES OF SOME FLAVONES,
CHALCONES AND 2,3-DIHYDROFLAVONES

Class of compound	Substituents	*Carbonyl stretching frequency (cm^{-1})
Flavone	—	1649
	3-OH	1619
	3-OMe	1648
	5-OH	1652
	7-OMe	1640
	3,5-OH	1638
2,3-Dihydroflavone	—	1695
	7-OMe	1685
	5-OH	1648
Chalcone	—	1659
	2′,3,4-OH	1621

* Spectra obtained in carbon tetrachloride solution[65, 81, 82].

position causes a decrease of approximately 30 cm^{-1} in the position of the carbonyl absorption and is presumably due to hydrogen bonding (26). A hydroxyl group in the 5 position has correspondingly little effect although the presence of a strongly hydrogen bonded hydroxyl group is readily discerned in the hydroxyl stretching frequency range and in this respect therefore 5-hydroxyflavones differ from other peri-hydroxy compounds. Suspected changes in the normal carbonyl stretching frequency due to hydrogen bonding may be confirmed by examination of the corresponding methylated derivatives. Removal of the chelation usually results in an increase in the carbonyl stretching frequency to a value approximately the same as that of the parent unsubstituted flavone. Hydroxyl and methoxyl groups in other parts of the flavone nucleus have only marginal effects on the carbonyl absorption but with 2,3-dihydroflavones, in which the carbonyl absorption is close to that of an aromatic ketone, substituents have a more pronounced effect. A hydroxyl group in the 5 position causes a considerable lowering (~ 47 cm^{-1}) which is probably due to chelation and the shift (~ 10 cm^{-1}) occurring with a hydroxyl or methoxyl group in position 7 can be ascribed to contributions to the resonance hybrid of structures such as (27).

The infra-red spectra of hydroxyflavan-3-ols and hydroxyflavan-3,4-diols are characterized by the absence of carbonyl absorption. The spectra of their phenolic methyl ethers, obtained by the action of diazomethane, confirms the presence of aliphatic hydroxyl groups generally in a hydrogen bonded form. On the basis of comparisons with model compounds Philbin and Wheeler[29] have assigned the hydroxyl stretching frequencies in the *cis*- and *trans*-flavan-3,4-diols (28 and 29) to OH$\cdots$O and OH$\cdots\pi$ hydrogen bonded species. The 3,4-*cis* configuration was attributed to the compound showing a doublet hydroxyl stretching frequency, that at 3606 cm^{-1} to OH$\cdots\pi$ bonding and that at 3578 cm^{-1} to OH$\cdots$O bonding. The single hydroxyl frequency in the 3,4-*trans* compound was interpreted as due to an overlap of the OH$\cdots\pi$ and OH$\cdots$O absorptions, the latter type of bonding being weaker in the *trans* configuration. Other workers[26] have suggested that the aliphatic hydroxyl group in the flavan-3-ols (+)-catechin and (−)-epicatechin tetramethyl ethers is axially disposed on the chroman ring

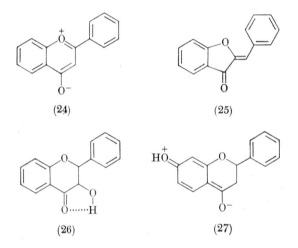

(24)   (25)

(26)   (27)

and hydrogen bonded to the ethereal oxygen atom at position 1 (8 and 9). Thus at least two interpretations of the hydrogen bonding associated with the 3-hydroxyl group in flavan-3-ols and flavan-3,4-diols have been put forward. The proton resonance spectra of flavan-3,4-diols favour the OH$\cdots\pi$ bonding since they show the 2-hydrogen to be axial and hence the 2-aryl group must be in an equatorial position, but the evidence is not unequivocal in the flavan-3-ol series[78].

Isoprenoid units in their various forms (e.g. 30, 31) when located in flavonoid compounds show characteristic infra-red absorption which assist in their identification[83]. Compounds bearing a $\gamma\gamma$-dimethylallyl substituent on an aromatic ring (30) show infra-red absorption characteristics typical of a trisubstituted ethylene with maxima in the range 790–840 cm^{-1} and of a gem dimethyl group (ν_{max} 1385–1380 cm^{-1}). The 2,2-dimethylchromene residue (31) similarly shows absorption characteristic of the gem-dimethyl and ethylenic groups (ν_{max} 1365, 1375 and 1590 cm^{-1}).

Further characterization of flavonoid materials is possible from their ultra-violet spectra. Extensive compilations[64] of spectral data concerning the various types of flavonoids makes a comparison with known compounds readily achieved. Flavones (including flavonols) show two principal absorption bands (I $\sim$ 240–270 mμ) and (II $\sim$ 320–380 mμ) the position, fine structure and intensity of which vary according to the substituents present in rings A and B (Table VIII). Attempts to empirically correlate the position and intensity of these absorption

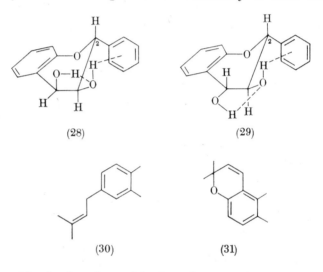

(28)   (29)

(30)   (31)

maxima with the location of hydroxyl or methoxyl groups in the flavone nucleus have led to suggestions that substituents in ring A predominantly influence the nature of band I whereas those in ring B effect band II. These deductions, which are based on the association of band I with the benzoyl grouping (32) and band II with the cinnamoyl group (33), undoubtedly represent too great a generalization of the facts as is shown by reference to Table VIII. Isoflavones and isoflavanones in contrast to the flavones show more intense absorption maxima for band I ($\sim$ 250–270 mμ) and lower for band II ($\sim$ 300–330 mμ). Afromosin (35) and other isoflavones which have the unusual 2,4,5-oxygenation pattern in ring A exhibit rather high intensity absorption in the 320–340 mμ range which has been attributed to this particular pattern of oxygenation[74]. Unexpectedly large bathochromic shifts in the ultra-violet absorption maxima of flavones and isoflavones may be

TABLE VIII

ABSORPTION MAXIMA AND EXTINCTION COEFFICIENTS OF SOME
FLAVONOID COMPOUNDS

Class	Substituents	λI_{max} (mμ)	log ε	λII_{max} (mμ)	log ε	
Flavone	—	250	4·07	297·5	4·20	
	3-OH	239 ⎱ 305 ⎰	4·14 3·86	347·5	4·04	
	4'-OH	251		327		
	5-OH	272	4·35	337	3·88	
	7-OH	245	4·17	345	4·28	
	5,7-OH	270	4·42	330	3·90	Chrysin
	5,7,4'-OH	265	4·25	340	4·31	Apigenin
	5,7,4'-OMe	265	4·25	325	4·33	Acacetin
	5,7,3',4'-OH	258	4·22	355	4·28	Luteolin
	3,7,3',4'-OH	252·5 ⎱ 315 ⎰	4·33 4·22	370	4·43	Fisetin
Flavanone	—	250	3·86	320	3·37	
	7,4'-OH	234, 237		312		Liquirntigenin
	5,7-OH	288	4·35	314	3·78	Pinocembrin
Isoflavone	—	245	4·41	307	3·82	
	7,4'-OH	232* ⎱ 250 260* ⎰	4·32	302	3·90	Dadzein
	5,7,4'-OH	263	4·57	325	3·71	Genistein
Chalcone	—	228	3·91	309·5	4·35	
	2',4,4'-OH	242		370		
	2',3,4,4'-OH	263	3·99	382	4·44	Butein
Aurone	—	251 ⎱ 316·5 ⎰	4·10 4·27	379	4·06	
	4-OH	225* ⎱ 307 ⎰	4·14 4·26	389	4·25	
	3',4'-OH	259 ⎱ 277 ⎰	4·09 4·26	330 ⎱ 415·5 ⎰	3·87 4·43	
Flavan-3-ol	5,7,3',4'-OH	280				Catechin

* Inflexion. Data compiled from 64, 84.

associated with conjugation to an external double bond such as in a
dimethyl chromene residue [e.g. ichthynone (19)] and in these cases
comparison of the dihydroderivative with model compounds gives a
more satisfactory result[74]. In the case of 2,3-dihydroflavones, where
ring B is not conjugated to the carbonyl group, the extinction coeffi-
cient associated with band II is considerably diminished. Polyhydroxy-
aurones and chalcones are yellow in colour and show strong absorption

at wavelengths in the region of 400 mμ, aurones are further distin-guished by the complexity and fine structure of their spectra[84]. The ultra-violet spectra of flavan-3-ols and flavan-3,4-diols are in contrast relatively simple and show a single absorption maxima at about 280 mμ typical of simple phenolic compounds.

The position of hydroxyl substitution in the flavone nucleus may to some extent be determined by utilizing the effect of various reagents on its chromophoric system[64]. Thus flavonoid compounds with a hydroxyl group suitably oriented with respect to the carbonyl group readily form metal complexes (e.g. 34). Aluminium chloride gives stable yellow complexes with 3- and 5-hydroxyflavones the absorption

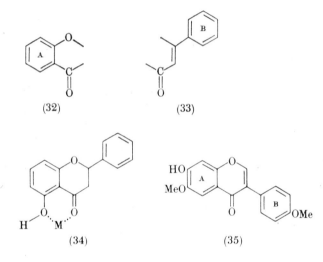

(32)   (33)

(34)   (35)

spectra of which show that the characteristic maxima (I and II) have undergone bathochromic shifts of the order 20–40 mμ, and in some cases have revealed a fine structure. Sodium acetate solution is sufficiently basic to ionize hydroxyl groups in the 3, 7 or 4' positions of flavone and results in changes in the absorption spectra. A characteristic batho-chromic shift (8–20 mμ) occurs in band I due to ionization of a 7-hydroxyl group and a similar change of 15–30 mμ occurs in band II by ionization of the 4'-hydroxyl but if a 3-hydroxyl group is present this latter effect is nullified. o-Dihydroxy and vicinal-trihydroxy groups in ring B produce a shift to higher wavelength (~ 15–30 mμ) of band II when treated with buffered boric acid solutions and this reagent pro-vides a sensitive test for hydroxyl groups with this particular orienta-tion.

Proton magnetic resonance spectroscopy has proved to be a particularly useful tool in the structure determination of flavonoid compounds since in general they contain only a few protons with a variety of electronic environments. The method has in particular provided useful information about the pattern of substitution in the aromatic nuclei, the presence of methoxyl, methylenedioxy and hydroxyl functions and the location of the various terpenoid residues in the more complex flavonoids[66]. The stereochemical problems posed by the rotenoids and the flavan-3,4-diols have also been satisfactorily elucidated by use of this technique. Many flavonoids, especially the glycosides, due to the presence of numerous hydroxyl groups have low solubility in the normal solvents suitable for proton magnetic resonance studies. Although to some extent this difficulty may be overcome by conversion to the methyl ethers in these cases naturally occurring methoxyl groups are then indistinguishable from the synthetic ones in the spectrum. The

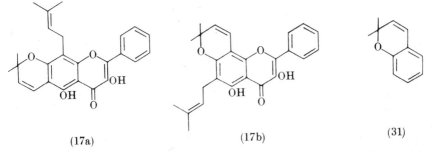

(17a)   (17b)   (31)

corresponding trimethylsilyl ethers do, however, appear to hold several advantages in this respect[85], thus they are readily and quantitatively formed (heating at 80° in anhydrous pyridine for two hours with hexamethyldisilazane), soluble in carbon tetrachloride, thermally stable and may be purified as required by vapour phase chromatography. Heating in 50% aqueous methanol under reflux for two hours hydrolyses the silyl ether back to the parent flavonoid compound which may be recovered unchanged. Unlike other methyl ethers the trimethylsilyl group absorbs at high fields and therefore does not interfere with other aliphatic protons if they are present in the flavonoid nucleus. Other workers have overcome solubility difficulties associated with the more hydrophilic flavonoids by carrying out the proton magnetic resonance measurements in deuterated dimethylsulphoxide[66].

Many correlations of the proton magnetic resonance spectra of flavonoids and related compounds have been made [66, 78] and in certain cases complete structural assignments are possible using this method of analysis. The work[71] which has led to the formulation of the flavonol sericetin (17a or b), isolated from the bark of *Mundulea sericeae*, provides a good example of the use of proton magnetic resonance spectroscopy in this field and also of the judicious combination of physical and chemical methods in structure elucidation. Sericetin ($C_{25}H_{24}O_5$) had a molecular weight of 404 derived from its mass spectrum. Ultra-violet and infra-red measurements showed sericetin to have a conjugated carbonyl group and two phenolic groups one of which was strongly chelated since the compound only readily formed monomethyl and monoacetyl derivatives. The pH dependence of the ultra-violet spectra of sericetin and its derivatives combined with the Shinoda colour test permitted its formulation as a 3,5-dihydroxyflavone with an ethereal oxygen function at position 7. Ring B of the flavone was shown to be an unsubstituted phenyl group by alkali fusion which gave benzoic acid. Biogenetic arguments led to the consideration of the additional $C_{10}H_{17}$ residue in sericetin as two isoprene units and this received

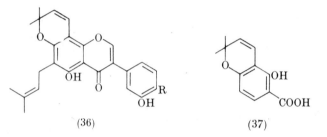

(36)   (37)

support from the Kuhn–Roth oxidation which gave 1·25 equivalents of acetic acid, the infra-red spectrum which indicated a gem dimethyl group (1375, 1361 cm^{-1}) and a conjugated olefinic double bond (1630 cm^{-1}) and the ultra-violet spectrum in the range 225–230 mμ which was typical of a 2,2-dimethylchromene (31). The formulation of sericetin as (17a or b) was supported by examination of the proton magnetic resonance spectrum and its comparison with those of osajin (36, R = H) and pomiferin (36, R = OH) and β-tubaic acid (37). The spectrum contained a complex band system (τ 2·57, 1·85) characteristic of a monosubstituted phenyl group (ring B) whose integrated peak area confirmed its association with five protons. A temperature sensitive band (τ 3·20) and a sharp line (τ −1·87) were attributed to the 3-hydroxyl and hydrogen bonded 5-hydroxyl group respectively and the remaining features of the spectrum due to the isoprene residues were interpreted as shown in Table IX by reference to the known compounds (36, R = H and R = OH) and (37). A chemical degradation to distinguish the alternative structures (17a or b) has not been reported.

Proton magnetic resonance studies have also been used to elucidate the stereochemistry of various flavonoids in particular the rotenoids[79]

TABLE IX

PROTON RESONANCE SPECTRUM OF SERICETIN IN
DEUTEROCHLOROFORM[71]

Group	Me\C/Me	Me\C=/Me	CH₂\C=C/	H\C=C/H	H\C=C/
τ-Value	8·55	8·33, 8·18	6·52	4·43	3·30
Number of protons	6	3 3	2	1	1

and flavan-3,4-diols[75–78] and its application in this latter case is discussed later.

Chemical methods have provided considerable insight into the structure and reactivity of the supposed precursors of condensed tannins—the flavan-3-ols and flavan-3,4-diols—and this aspect of their properties is discussed below. Suitable methods for the degradation of the other co-occurring flavonoids have been adquately enumerated elsewhere.

VI. REACTIONS AND STEREOCHEMISTRY OF FLAVAN-3-OLS

The chemical reactions which establish the structure of a compound as a flavan-3-ol and provide information regarding the relative configuration of the 2-aryl and 3-hydroxyl groups are illustrated by reference to Figs. 3 and 4. Shown in Fig. 3 are some of the reactions used[86] to characterize (−)-epiafzelechin (38), a flavan-3-ol isolated by King and his colleagues from the heartwood of *Afzelia* species, and in Fig. 4 some analogous reactions of (+)-catechin (45, $R^1 = R^2 = H$) (*Uncaria gambir* and other plant species) are depicted. Epiafzelechin formed a tetra-acetate and with dimethyl sulphate and potassium carbonate a tri-*O*-methyl ether which on oxidation with potassium permangate gave anisic acid. The tri-*O*-methyl ether gave a monoacetate and *p*-toluene sulphonate (39) and the latter on treatment with hydrazine underwent a smooth base catalysed elimination of *p*-toluene sulphonic acid to give the flav-2-ene (40). This reaction, which has been interpreted as a stereospecific *trans*-E_2-elimination in which the hydrogen at position 2 and the *p*-toluenesulphonyl group at position 3 have the *trans*-diaxial relationship (39), thus established the *cis* disposition of the 2-aryl and 3-hydroxyl groups in epiafzelechin and showed it to be a member of the "epi" series. Application of this particular reaction to flavan-3-ols with a *trans* configuration of the 3-hydroxyl and 2-aryl groups, e.g. (+)-catechin (45, $R^1 = R^2 = H$), results in fragmentation of the molecule. The characteristic stereochemistry of these compounds

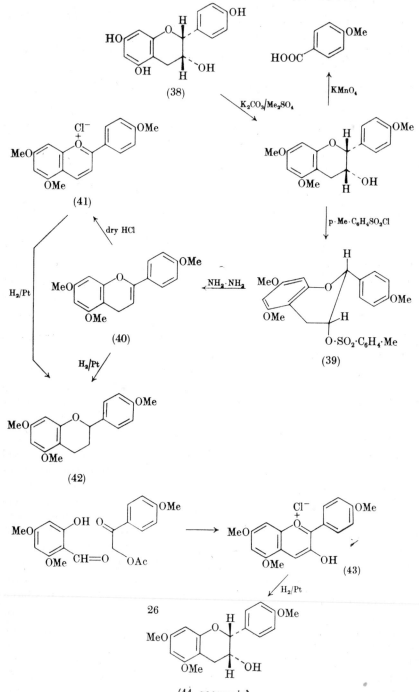

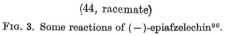

Fig. 3. Some reactions of (−)-epiafzelechin[86].

is, however, readily illustrated[28] by the facile 1, 2-shifts, involving neighbouring group participation of the 2-aryl group, which they undergo. Thus the tetramethyl ether (45; R^1 = Me, R^2 = H) when treated with phosphorus pentachloride gave[87] the 2-chloro-isoflavan

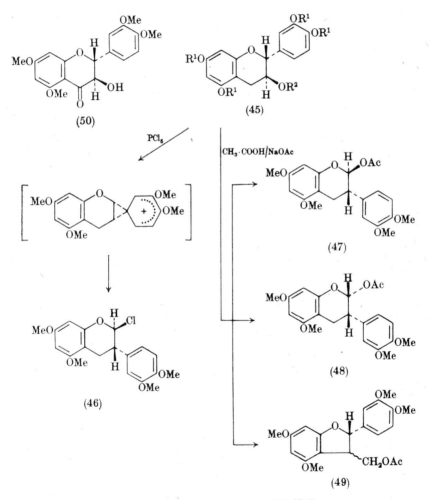

FIG. 4. Some reactions of (+)-catechin[87-89].

(46) and similarly acetolysis of the p-toluene sulphonate (45; R^1 = Me, R^2 = —$SO_2C_6H_4Me$) gave the diastereoisomeric 2-acetoxy-iso-flavans (47 and 48)[28]. It has been shown[88] that the absolute configuration of the 2-chloro-isoflavan is as shown in (46) and with the configuration

of (+)-catechin also as illustrated (45; $R^1 = R^2 = H$) this indicates that the rearrangement follows the normal stereochemical course of a Wagner–Meerwien change. In addition to the two epimeric acetates obtained by acetolysis of tetra-O-methyl (+)-catechin toluene-p-sulphonate (45; $R^1 = Me$, $R^2 = —SO_2C_6H_4Me$) a third product (49) was isolated[89] and its production most probably proceeds by migration of the phenyl ring A of the catechin system from C_2 to C_3. Whalley[28] has suggested that the formation of the epimeric acetates (47) and (48) in the same reaction probably means that the acetolysis is not completely synchronous.

With anhydrous hydrogen chloride the flav-2-ene (40) from (−)-epiafzelechin gave the tri-O-methyl ether of apigenidin (41)[90] and reduction of apigenidin (41) or the flav-2-ene gave the flavan (42). Final confirmation of the structure and relative stereochemistry of the groups in (−)-epiafzelechin (38) was obtained by a synthesis of (±)-epiafzelechin trimethyl ether. Reduction of 5,7,4′-trimethylpelargonidin chloride (43) according to the method of Freudenberg[91] proceeded by *cis* addition of hydrogen to the planar flavylium system and gave a racemate (44) identical in its infra-red spectrum with the natural product. Synthesis of racemic flavan-3-ols of the catechin series with a 2,3-*trans* orientation of the 2-aryl and 3-hydroxy groups may be achieved by reduction of the corresponding 2,3-*trans*-dihydroflavonol. Thus reduction of (+)-taxifolin tetramethyl ether (50) gave[92] (+)-catechin tetramethyl ether (45; $R^1 = Me$, $R^2 = H$). The racemic 2,3-*trans*-dihydroflavonols may be prepared by a variety of methods[93] and these serve as readily available starting materials for the synthesis of racemic flavan-3-ols of the same configuration at C_2 and C_3. Thus the action of water or alkali on 2′-hydroxychalcone dibromides and the alkaline peroxide oxidation of 2′-hydroxychalcones have both been formulated[94, 95] as proceeding via the epoxide (51) to the 2,3-*trans*-dihydroflavonols (52).

Further corroborative evidence for the stereochemical configuration at C_2 and C_3 in (−)-epiafzelechin was obtained by a comparison of the molecular rotations of the natural product, its tri-O-methyl ether and the latter's acetate and p-toluene sulphonate with the corresponding derivatives of (−)-epicatechin and (+)-catechin (Table X). These clearly implied an identical relationship between the optically active centres of (−)-epiafzelechin and (−)-epicatechin and established the former compound as a member of the (−)-epi series. This comparative method has also been used[96] to infer the stereochemical configurations of other flavan-3-ols and in this way (+)-gallocatechin (53; $R^1 = R^2 = OH$), (−)-robinetinidol (53; $R^1 = H$, $R^2 = OH$) and (−)-fisetinidol (53; $R^1 = R^2 = H$) have been related to the (+)-catechin (53;

R^1 = OH, R^2 = H) series and $(-)$-epigallocatechin (54; (R = OH) to $(-)$-epicatechin (54; R = H).

The absolute configuration at the asymmetric centres C_2 and C_3 of $(+)$-catechin (53; R^1 = OH, R^2 = H) and $(-)$-epicatechin (54; R = H)

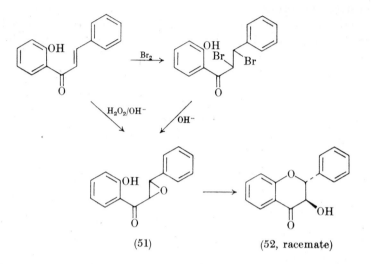

(51)   (52, racemate)

and several other flavan-3-ols have been established by Prelog's atro-lactic acid method[97] and also by ozonolysis[98, 99]. Using the former method[26] $(-)$-epicatechin tetramethyl ether gave atrolactic acid with $[\alpha]_D$ $-$ 16·4° corresponding to 43% excess of the $(-)$-isomer (55). The absolute configuration of $(-)$-epicatechin followed therefore as (54;

TABLE X

COMPARISON OF THE MOLECULAR ROTATIONS (M) OF
$(-)$-EPIAFZELECHIN, $(-)$-EPICATECHIN AND
$(+)$-CATECHIN AND THEIR DERIVATIVES

Derivative	$(-)$-Epiafzelechin[a]	$(-)$-Epicatechin[b]	$(+)$-Catechin[b]
Natural product	$-$ 16,200	$-$ 20,000	0[c]
			$+$ 4930[d]
Methyl ether	$-$ 21,200	$-$ 21,300	$-$ 4640
Methyl ether acetate	$-$ 26,400	$-$ 27,600	$+$ 2640
Methyl ether p-toluenesulphonate	$-$ 4230	$-$ 8450	$+$ 11,400

Calculated from figures by King, Clark-Lewis and Forbes[86].
[a] $[M]_D^{20}$; [b] $[M]_{Hg}^{20}$; [c] ethanol; [d] Water–acetone (1:1).

R = H). Application of the atrolactic acid method to (+)-catechin tetramethyl ether resulted in a low degree of asymmetric synthesis but independent confirmation of the absolute configuration of (+)-catechin and also of (−)-epicatechin has been obtained by Hardegger using exhaustive ozonolysis[98, 99]. Under these conditions (+)-catechin (53; R^1 = OH, R^2 = OH) gave the acid (56) which when esterified and reduced with lithium aluminium hydride gave the alcohol (57) identical

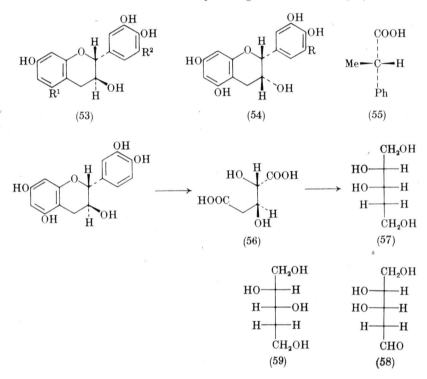

to that prepared by reduction of 2-deoxy-D-ribose (58). Under the same conditions (−)-epicatechin (54; R = H) furnished the isomeric 2-deoxy-D-xylitol (59). Using the nomenclature of Cahn, Ingold and Prelog[25] (+)-catechin (53; R^1 = OH, R^2 = H) and (−)-epicatechin (54; R = H) have therefore the 2R, 3S and 2R, 3R configurations respectively. The absolute configurations of the other naturally occurring flavan-3-ols so far isolated have been determined by application of similar methods or deduced from a comparison of molecular rotations and are shown in Table XI. A point of interest is the uniformity of absolute configuration (2R) at C_2 in the natural flavan-3-ols.

Flavan-3-ols undergo epimerizations readily in hot aqueous solution

TABLE XI

STRUCTURE AND CONFIGURATION OF SOME NATURALLY
OCCURRING FLAVAN-3-OLS

Flavan-3-ol	Substituents	Absolute configuration	Sources
(+)-Catechin	5,7,3',4'-OH	2R, 3S	Uncaria gambir[6], leaves Camellia sp.[54], bark Acacia[54, 55], Quercus[60] and Castanea[61] sp.
(−)-Epicatechin	5,7,3',4'-OH	2R, 3R	Acacia catechu[7], bark Acacia sp, leaves Camellia sp.
(+)-Gallocatechin	5,7,3',4',5'-OH	2R, 3S	leaves Camellia sp.[57], bark Acacia[54, 55], Quercus[60], Castanea[61] sp.
(−)-Epigallocatechin	5,7,3',4',5'-OH	2R, 3R	leaves Camellia[57] sp., bark Acacia sp.[54, 55]
(−)-Robinetinidol	7,3',4',5'-OH	2R, 3S	heartwood Robinia pseudacacia[52], bark Acacia sp.[54, 55]
(−)-Fisetinidol	7,3',4'-OH	2R, 3S	heartwood Acacia mollisima[100]
(−)-Epiafzelechin	5,7,4'-OH	2R, 3R	heartwood Afzelia sp.[86]
(+)-Afzelechin	5,7,4'-OH	2R, 3S	kino Eucalyptus calophylla[101]

and the relevant interconversions for (+)-catechin and (−)-epicatechin which have been summarized by Freudenberg[9, 48] are shown below:

(+)-catechin   → (+)-epicatechin (mainly)

(−)-epicatechin → (−)-catechin (mainly).

It is clear that inversion occurs at only one of the two optically active centres during these changes and Freudenberg assumed that this was at the 2 position. Birch and co-workers[26] proved this assumption correct by reductively removing the asymmetry at the 2-position in the tetra-methyl ethers of (+)-catechin and (−)-epicatechin using sodium in liquid ammonia reduction. Methylation of the resultant phenolic alcohols gave from (−)-epicatechin the diarylpropan-2-ol (60) with an excess of the dextrorotary enantiomer and from (+)-catechin the same alcohol containing an excess of the laevorotary form (61), further confirming that (+)-catechin and (−)-epicatechin have opposite configurations at C_3. The epimerization of (+)-catechin and (−)-epicatechin must consequently involve inversion at the 2 position in the heterocyclic ring. Flavan-3-ols which lack a hydroxyl group in the 4'

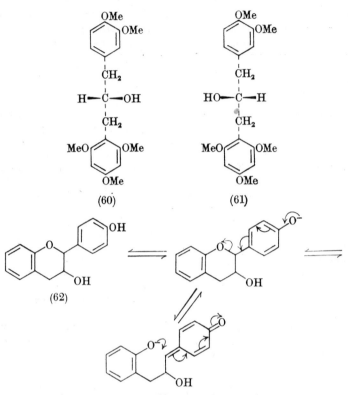

(60)   (61)

(62)

(63)

position are stable to epimerization and Whalley[28] on this basis has suggested the process occurs as shown (62 → 63).

VII. Reactions and Stereochemistry of Flavan-3,4-diols

The most characteristic reaction of flavan-3,4-diols is undoubtedly their partial conversion[13-15] with hot acid to the corresponding anthocyanidins. Further reactions of flavan-3,4-diols which are employed in the detailed elucidation of the structure and stereochemical relationships of these compounds are illustrated for (−)-melacacidin in Fig. 5. King and his co-workers[102, 103] isolated (−)-melacacidin (64) from the heartwood of *Acacia melanoxylon* and Clark-Lewis and Mortimer[104] later isolated the same compound in a crystalline state from other *Acacia* species. Treated with diazomethane it gave a tetramethyl ether (65) which itself gave a diacetate. Oxidation of the tetramethyl ether with potassium permanganate produced 2-hydroxy-3,4-dimethoxy and 3,4-dimethoxybenzoic acids from rings A and B respectively. King and

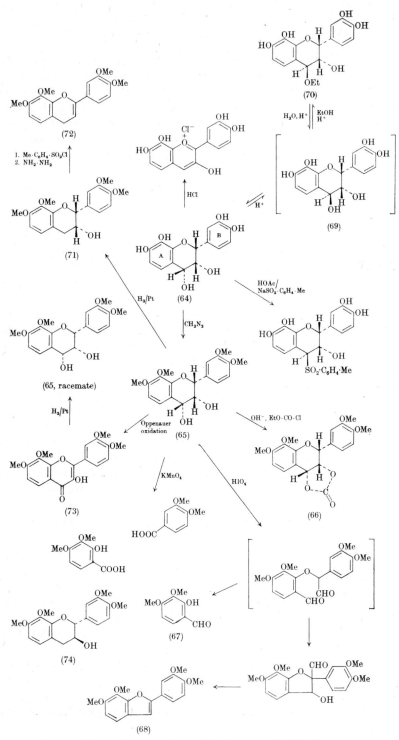

FIG. 5. Some reactions of (−)-melacacidin[102-105].

Bottomley[102] interpreted the formation of the cyclic carbonate (65) when the tetramethyl ether was treated with ethyl chloroformate and alkali as indicating the presence of a cis-3,4-diol grouping in the natural product. Confirmation of this assignment was obtained by the ready cleavage of the diol grouping in the tetramethyl ether with periodic acid to give 2-hydroxy-3,4-dimethoxybenzaldehyde (67) and the benzofuran (68)[105] which it was suggested probably arose from the initial product of reaction by the pathway shown. The analogous reaction of flavan-3,4-diols with lead tetra-acetate has been investigated by Brown, Bokadia and Cummings[106] with similar results.

Associated with (−)-melacacidin in extracts of *Acacia melanoxylon* are small amounts of an isomer isomelacacidin (69). Heating the crude extracts with ethanol converted isomelacacidin to its 4-O-ethyl derivative (70) which was readily separated from (−)-melacacidin by countercurrent distribution[104]. Isomelacacidin was readily produced from (−)-melacacidin by heating with dilute acid and Clark-Lewis and Mortimer suggested that the two compounds were epimers at the 4 position. A similar isomerization has been described for the related (−)-teracacidin and isoteracacidin which are isolated from the heartwood of *Acacia intertexta*[107].

Elucidation of the stereochemistry of flavan-3,4-diols has presented subtle and elusive problems particularly in relation to the *cis* or *trans* configurations of the 3,4-diol grouping. The stereochemistry at the 2 and 3 positions of the heterocyclic ring is most conveniently determined[108] by reductive removal of the oxygen function at position four to give the corresponding flavan-3-ol which may then be classified as belonging to the catechin or epicatechin series by comparison of the molecular rotations or optical rotary dispersion curves. Hydrogenation of (−)-melacacidin tetramethyl ether (65) gave[109] the tetramethoxy flavan-3-ol (71) whose initially high laevo rotation was enhanced on conversion to the 3-acetate but depressed on formation of the 3-tosylate suggesting that it was an analogue of (−)-epicatechin (see Table X) with the cis-2R,3R-configuration. This fact was confirmed by the observation that the 3-tosylate of (71) underwent a smooth base catalysed elimination of p-toluene sulphonic acid to give the flav-2-ene (72). Combined with the fact that (−)-melacacidin underwent ready formation of a cyclic carbonate (66) and contained therefore a cis-3,4 diol grouping the natural product was, on this evidence, assigned[109] the all cis configuration (2R, 3R, 4R). Unequivocal proof of this configuration was obtained[103] by synthesis of the racemates of melacacidin tetramethyl ether, of the 2,3-cis-flavan-3-ol (71) and of the corresponding 2,3-trans-flavan-3-ol (74). Raney-nickel hydrogenation of the flavonol (73) (prepared by Oppenauer oxidation of ($-$)-melacacidin tetramethyl ether (65)) gave ($\pm$)-tetramethyl melacacidin identical in its infra-red spectrum with the tetramethylether of the natural product. The reduction of (73) was assumed to proceed by an all *cis* addition to the heterocyclic ring by analogy with other reductions (e.g. cyanidin →　($\pm$)-epicatechin). An authentic racemate of the 2,3-*cis*-flavan-3-ol (71) resulted from further hydrogenation of ($\pm$)-melacacidin tetramethyl-ether over a palladium catalyst and differed considerably in its infra-red spectrum from the related 2,3-*trans*-flavan-3-ol (74) which was

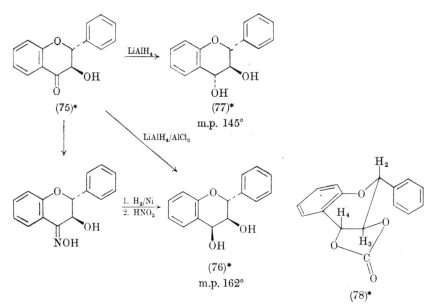

* 75, 76, 77, 78 all racemates, only the 2R enantiomers shown.

prepared by a two-stage reduction of ($\pm$)-7,8,3',4'-tetramethoxy-2,3-*trans*-dihydroflavonol.

The observations of Brown and co-workers[110] and Corey, Philbin and Wheeler[77] on the formation of cyclic derivatives of flavan-3,4-diols have shown that the use of such methods can, however, lead to erroneous formulations of the stereochemistry of the 3,4-diol grouping. Reduction of 2,3-*trans*-3-hydroxyflavan-4-one (75) with lithium aluminium hydride gave a flavan-3,4-diol (m.p. 145°); reduction with lithium aluminium hydride in the presence of aluminium chloride[110] yielded a further diol (m.p. 162–3°) which was also identical with the flavan-3,4-diol prepared from the oxime of (75) using the method of Bognar, Rakosi, Fletcher, Philbin and Wheeler[111]. Both flavan-3,4-diols gave cyclic carbonate derivatives with ethyl chloroformate and gave the same isopropylidene derivative with acetone and anhydrous hydrogen chloride; the diol (m.p. 162–3°) in 70% yield and the diol (m.p. 145°) in 10% yield. Hydrolysis of the isopropylidene derivative gave in both cases the same diol (m.p. 162°). Brown and his colleagues[110] concluded from this evidence that the diol (m.p. 162°) was the *cis*-diol (76) and that the one with m.p. 145° was the *trans*-diol (77). Measurements[77] of the proton magnetic resonance of the cyclic carbonates of (76) and (77) supported these conclusions. The spectrum of the carbonate of the *trans*-diol (78) was unexpectedly simple showing peaks due to aromatic protons (integrated intensity 9) and at higher field the heterocyclic protons which appeared as a doublet (intensity 2) and a triplet (intensity 1). The only tenable explanation of this spectrum was that the carbonate had the conformation (78) with the heterocyclic protons in the axial *trans* positions; those at 2 and 4 having the same chemical shift and coupling constant (10·6 c/s) relative to that at the 3 position giving rise to the doublet (intensity 2) and that at position 3 giving the triplet (intensity 1). This work demonstrated therefore that cyclic carbonate formation may take place with a *cis*- or *trans*-3,4-diol and that a *trans*-3,4-diol may under the conditions of reaction undergo inversion at the 4 position and form the isopropylidene derivative of the corresponding *cis*-3,4-diol. Methods for the determination of the relative configuration of the diol grouping in flavan-3,4-diols by formation of cyclic derivatives are seen therefore to be equivocal and a more reliable chemical method appears to be measurement of the rate of oxidation of the diol with lead tetra-acetate or periodic acid[112, 113]. Brown and his co-workers observed[106] that the *cis*-diols were more readily cleaved by oxidizing agents such as lead tetra-acetate than the corresponding *trans*-compounds and Clark-Lewis and Williams[112] and Roux and Drewes[113] have extended this observation to a variety of other diols. The rate of reaction for the periodate oxidation of a number of compounds was claimed by Roux and Drewes[113] to be six to nine times as fast for the 3,4-*cis*- as for the 3,4-*trans*-isomers and, with suitable standards, measurement of the rate of oxidation may therefore serve as an indication of the relative configuration of the 3,4-diol grouping. Clark-Lewis and Williams[112, 78], however, suggested that in the case of 2,3-*trans*-flavan-3,4-diols the rates of oxidation did not permit a clear distinction between the 3,4-*cis*- and *trans*-configurations and they indicated that the method was not wholly reliable in such instances.

The work of Clark-Lewis, Jackman and Williams[75, 76, 78] and later Roux and Drewes[113] has shown that the stereochemistry of flavan-3,4-diols may be unambiguously and most easily elucidated from measurements of their proton magnetic resonance spectra. The work of Clark-Lewis and his colleagues[75, 76] is quoted in some detail here since it also permits a brief description of the methods available for the synthesis of the four possible racemates of the flavan-3,4-diol structure. The methods employed by these workers were similar in many respects to those used by Kulkarni, Joshi and Kashikar[114, 115] for the synthesis of the racemates of 6-methyl-4'-methoxyflavan-3,4-diol, but some of the conclusions relating to the configuration of the 2,3-*trans*-flavan-3,4-diols of this series have been subject to revision in later work since they were based on the formation of cyclic derivatives.

The syntheses of the four racemates of the 6-methyl-3',4'-dimethoxy-flavan-3,4-diol structure by Clark-Lewis, Jackman and Williams[76] are summarized in Fig. 6. Hydrogenation of flavonols is the only route developed so far for the preparation of 2,3-*cis*-flavan-3,4-*cis*-diols and application of this method gave the racemate (79). Dependent on the conditions used reduction of 2,3-*trans*-dihydroflavonols may be used to give either the corresponding 3,4-*cis*- or 3,4-*trans*-diol. Thus reduction with lithium aluminium hydride or hydrogenation over platinum charcoal of (80) gave the 2,3-*trans*-flavan-3,4-*trans*-diol (81) but reduction with lithium aluminium hydride–aluminium chloride according to the procedure of Brown[110] gave the 2,3-*trans*-flavan-3,4-*cis*-diol (82). The remaining racemate, the 2,3-*cis*-flavan-3,4-*trans*-diol (83), was the least accessible* but was prepared according to the method of Kulkarni and Joshi[114] from the 3-bromoflavanone (84). The configuration assigned to the flavan-3,4-diol (83) followed mainly from the fact that it differed from the other racemates (79), (81) and (82) since the stereochemistry of the intermediates in this reaction remain poorly defined.

Proton magnetic resonance studies constitute one of the most powerful methods now available for the analysis of the configuration and conformation of substituted six-membered ring compounds[116]. The stereochemistry can be investigated because the coupling constant between the two hydrogen atoms in the H—C—C—H system depends on the dihedral angle between the two C—H bonds; being at a maximum when this is at 180° (i.e. a *trans*-conformation) and at a minimum for approximately 90° (i.e. *gauche*-conformation). It is also recognized, however, that in addition to angle dependency of coupling constants there exist superimposed substituent effects whose magnitudes are

---

* A novel and improved synthesis of the diacetates has recently been announced (J. W. Clark-Lewis and L. R. Williams, *Austral. J. Chem.* **18**, 90 (1965)). Acetylation of the 2,3-*cis*-flavan-3,4-*cis*-diol by boiling with acetic acid, acetic anhydride and potassium acetate proceeds in certain cases with inversion to give the corresponding 2,3-*cis*-flavan-3,4-*trans*-diol diacetate. The reaction has been postulated as proceeding via a cyclic *ortho*-ester form but its scope has not been thoroughly investigated.

difficult to predict. Although these effects are usually not significant enough to interfere in the differentiation between axial–axial and axial–equatorial or equatorial–equatorial orientation of hydrogen atoms on adjacent carbons they undoubtedly contribute to the considerable variations in the reported axial–axial (J_{aa}) coupling constants and axial–equatorial (J_{ae}) coupling constants for cyclic systems, for which

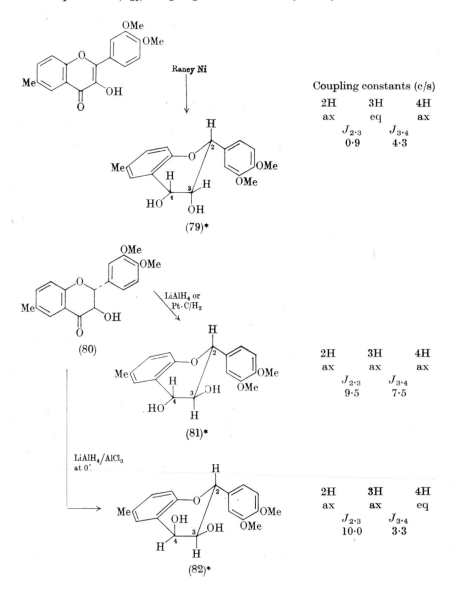

Coupling constants (c/s)

2H ax	3H eq	4H ax
	$J_{2\cdot3}$ 0·9	$J_{3\cdot4}$ 4·3

(79)*

2H ax	3H ax	4H ax
	$J_{2\cdot3}$ 9·5	$J_{3\cdot4}$ 7·5

(81)*

2H ax	3H ax	4H eq
	$J_{2\cdot3}$ 10·0	$J_{3\cdot4}$ 3·3

(82)*

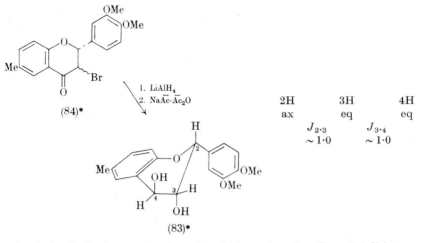

	2H	3H	4H
	ax	eq	eq
		$J_{2 \cdot 3}$	$J_{3 \cdot 4}$
		$\sim 1 \cdot 0$	$\sim 1 \cdot 0$

FIG. 6. Synthesis of racemates of the 6-methyl-3′,4′-dimethoxyflavan-3,4-diol structure, and coupling constants of protons attached to heterocyclic ring.

* All structures racemates, only 2R enantiomers shown.

the theoretical treatment of Karplus[117] predicts 9·2 and 1·72 c/s
respectively. Measurement of the proton magnetic resonance spectra
of the diacetates of (79), (81), (82), (83) confirmed[76] the configurations
shown and indicated that the heterocyclic ring adopts a conformation
close to the half chair form with the 2-hydrogen axial and therefore the
2-aryl group equatorial in each case. The relative stereochemistry at
both the 2, 3 and 3, 4 positions was clearly defined by the spin–spin
coupling constants. In the 2,3-*cis*-compounds $J_{2 \cdot 3}$ (or J_{ae}) was 0·9–
1·0 c/s whereas for the 2,3-*trans*-isomers $J_{2 \cdot 3}$ (or J_{aa}) was 9·5–10 c/s
and a similar clear distinction was recorded in the spin–spin coupling
constants for the 3 and 4 protons, although in this case the observed
variation from the expected values for the various coupling constants
may be due to substituent effects. Similar values for the coupling
constants for the 2, 3 and 4 protons of derivatives of other 2,3-*trans*-
flavan-3,4-diols have been recorded in later work by Roux and
Drewes[113].

Brown and his colleagues[110] have interpreted the stereochemistry of
the acetates of the flavan-3,4-diols (76) and (77) on the basis of the
τ-values for the methyl (acetyl) proton resonance peak, following
Lemieux, Kullnig and Moir[118] who observed in earlier work with
acetylated pyranose sugars that the peaks for equatorial acetoxy
groups occurred at higher field than those for axial acetoxy groups.
Thus the acetates of (76) and (77) showed acetate τ-values of 8·16 which

were attributed in both cases to the equatorial 3-acetoxy group. The second acetate peak of (77) at τ 8·02 was at considerably higher field than that of the second acetoxy group of (76) at τ 7·87. Hence the 3,4-*trans*-configuration was adopted for (77) with both acetoxy groups in equatorial positions and the 3,4-*cis*-configuration for (76) with the 3-acetoxy group in an equatorial and the 4 in an axial position. Jackman, Clark-Lewis and Williams[76, 78] quote τ-values for the acetate peaks of (79, 81, 82) in agreement with the above deductions but those for (83) constitute an apparent exception to this interpretation. Clark-Lewis[78] has indicated, however, that in view of the powerful shielding influence of the aromatic nuclei on the flavan system there is no *a priori* reason for expecting the Lemieux correlations[118] to apply to the flavan system; the validity of this approach is therefore questionable.

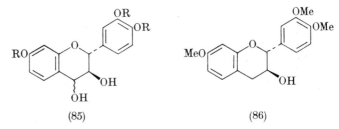

(85)   (86)

Clark-Lewis and Katekar[119] subsequently applied nuclear magnetic resonance measurements to revise the accepted configuration of (+)-mollisacacidin (85, R = H) a flavan-3,4-diol from *Acacia mollisima*. Hydrogenation of (+)-mollisacacidin trimethylether gave[109] (−)-fisetinidol trimethylether (86) which molecular rotation comparisons showed was an analogue of (+)-catechin (45, $R^1 = R^2 = H$) and therefore to have the (2R, 3S) configuration. The formation of cyclic derivatives of (+)-mollisacacidin indicated originally[120] a 3,4-*cis*-diol and hence the absolute configuration (2R, 3S, 4S) for this compound. However, measurements of the proton magnetic resonance spectra of (+)-mollisacacidin and its derivatives and in particular the coupling constants for the 3 and 4 protons indicated a 3,4-*trans*-diol structure and hence modification of the absolute configuration to (2R, 3S, 4R). Once again the formation of cyclic derivatives of flavan-3,4-diols is seen therefore to be an unreliable criterion of the 3,4-*cis*-configuration. Roux and Drewes[113] have confirmed the assigned configuration of (+)-mollisacacidin by similar methods and have shown (+)-leucorobinetinidin to have the (2R, 3S, 4R) and (−)-leucofisetinidin the (2S, 3R, 4S) configuration. The absolute configurations of other natural flavan-3,4-diols are recorded in Table XII and it may be observed that the configuration 2S in (−)-leucofisetinidin is noteworthy as being antipodal to the majority of other naturally occurring compounds of this class.

The surveys of plant species by Robinson and Robinson[13] and by Bate-Smith[14, 15] indicated that the flavan-3,4-diols corresponding to delphinidin and cyanidin were most widely distributed in plant species. Although several leucodelphinidins and leucocyanidins have been isolated from natural sources their stereochemical configurations have not been determined and in several cases further structural characterization is also desirable. These and other flavan-3,4-diols in this category are summarized in Table XIII.

In addition to the flavan-3,4-diols listed in Tables XII and XIII several compounds have been isolated from Nature whose structures are closely related to the flavan-3,4-diols and may be considered to be

TABLE XII

STRUCTURE AND ABSOLUTE CONFIGURATION OF NATURALLY OCCURRING FLAVAN-3,4-DIOLS

Flavan-3,4-diol	Substituents	Absolute configuration	Source
(−)-Melacacidin	7,8,3′,4′-OH	2R, 3R, 4R	Heartwood of *Acacia melanoxylon, excelsa* and *harpophylla*[102–104]
(−)-Teracacidin	7,8,4′-OH	2R, 3R, 4R	Heartwood of *Acacia intertexta*[107]
(+)-Mollisacacidin			Heartwood of *Acacia mollisima*[120]
(+)-Gleditsin	7,3′,4′-OH	2R, 3S, 4R	*Gleditsia japonica*[121]
(−)-Leucofisetinidin	7,3′,4′-OH	2S, 3R, 4S	*Schinopsis lorentzii*[53], *Cotinus coggyria*[108]
(+)-Leucorobinetinidin	7,3′,4′,5′-OH	2R, 3S, 4R	Heartwood of *Robinia pseudacacia*

TABLE XIII

Flavan-3,4-diol	Substituents	Source
(+)-Leucocyanidin	5,7,3′,4′	Tamarind seed testa[122]
(+)-Leucocyanidin	5,7,3′,4′	*Butea frondosa* gum[123]
(−)-Leucopelargonidin	5,7,4′	*Eucalyptus calophylla* gum[124]
(−)-Leucodelphinidin	5,7,3′,4′,5′	*Cleistanthus collinus* (Karada)[125]
(+)-Leucodelphinidin	5,7,3′,4′,5′	*Eucalyptus pilularis* kino[126]
Leucodelphinidin	5,7,3′,4′,5′	Bark, *Quercus* sp.[60]
(+)-Guibortacacidin	7,4′	*Guibourtia coleosperma*[12]

3+

derived biogenetically from them. The structures established or suggested for peltogynol (87) from *Peltogyne porphyrocardia*[127], cyanomaclurin (88a, b or c) from *Artocarpus integrifolia*[128-130] and for the "proanthocyanidin" (89) from *Crataegus oxyacantha*[131] are interesting examples of this class of compound. Cacao leucocyanidin one of the phenolic constituents of fresh cacao beans[132] probably has a similar structure to (89) since on acid treatment it also yields cyanidin and

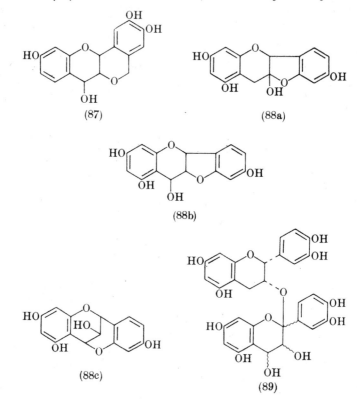

(−)-epicatechin. The structure suggested for the "proanthocyanidin" (89) (and the probable structure of cacao leucocyanidin) bears out the point that not all leucoanthocyanidins which have been revealed in the wide botanical surveys are necessarily simple flavan-3,4-diols.

VIII. Isolation and Identification of Stilbene Compounds

It is normally assumed that the condensed tannins from a wide variety of plant species have the flavan-3,4-diols and flavan-3-ols as their common precursors. Grassmann[18] has, however, put forward the theory that the tannins of spruce bark (*Picea abies*) are derived by polymerization of the stilbene piceatannol (5, R = OH) which either free or as glucosides is present in high concentration in the bark. Hydroxystilbenes have been isolated from other tannin bearing woods

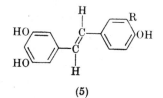

(5)

such as *Eucalyptus wandoo*[17] and it is possible that along with the flavan-3,4-diols and flavan-3-ols they act as precursors of condensed tannins in several woody species. In Table XIV a number of the more commonly occurring hydroxystilbenes and their plant sources are shown.

TABLE XIV

SOME NATURALLY OCCURRING HYDROXYSTILBENES

Compound	Substituents	Source
Pinosylvin	3,5-OH	*Pinus* species
Pinosylvin monomethyl ether	3-OH; 5-OMe	*Pinus* species[133]
Pterostilbene	3,5-OMe,4'-OH	*Pterocarpus santalinus*[134]
Resveratrol	3,5,4'-OH	*Veratrum grandiflorum*[135], *Eucalyptus wandoo*[17]
Hydroxyresveratrol	3,5,2',4'-OH	*Veratrum grandiflorum*[135], *Toxylon pomiferum*[136], *Artocarpus lakoocha*[137]
Piceatannol	3,5,3',4'-OH	*Vouacoupa macropetala*[138], *V. americana*[138], *Picea abies*[20]
Rhapontigenin	3,5,3'-OH,4'-OMe	Rhapontic spice[139]
Pentahydroxy stilbene	3,4,5,3',5'-OH	*Vouacoupa macropetala*[138] *Eucalyptus astringens*[140]

Paper chromatography again provides a useful way in which to characterize hydroxystilbenes in a plant extract. These compounds are readily located on chromatograms by their bright blue fluorescence in ultra-violet light and also by the application of spray reagents which are specific for the various orientations of phenolic hydroxyl groups.

Hydrogenation of a sample of the plant extract prior to paper chromatographic analysis converts the stilbenes to their dihydro derivatives with a resultant quenching of the fluorescence in ultra-violet light. A typical paper chromatogram of the ether soluble phenols of *Eucalyptus wandoo*[17] heartwood which contains 3,5,4'-trihydroxystilbene (5, R = H) and its 3-β-D-glucoside is shown in Fig. 7. The methods used for the isolation and identification of these compounds are typical for hydroxystilbenes and are described briefly below. The stilbene (Fig. 7(2)) and its 3-β-D-glucoside (Fig. 7(1)) were isolated by chromatography of the ether soluble phenols on cellulose powder using 0·5 N acetic acid as eluant and the glucoside was further purified by chromatography on perlon powder[17]. Both compounds showed ultra-violet absorption (λ_{max} 220 and 305 mμ) typical of stilbenes and had characteristic blue fluorescences in ultra-violet light which were quenched on reduction to

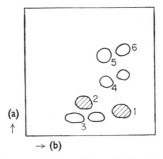

(a)
↑

→ (b)

FIG. 7. Paper chromatogram of the ether soluble phenols of *Eucalyptus wandoo*. Solvents: (a) 6% acetic acid; (b) butan-1-ol–acetic acid–water (6:1:2). Compounds identified: 1, resveratrol 3-β-D-glucoside; 2, resveratrol; 3, ellagic acid; 4, catechin; 5, chlorogenic acid; 6, p-coumaroylquinic acid.

the dihydro state. The free stilbene gave a triacetate and tri-O-methyl ether and was identical with a sample of 3,5,4'-trihydroxystilbene prepared as described by Takaoka[141]. The glucoside was hydrolysed by emulsin and was therefore a β-glucoside, on acetylation it gave a hexaacetate which when oxidized and the products de-acetylated yielded p-hydroxybenzoic and 3,5-dihydroxybenzoic acids. The attachment of the glucose at the 3-position was determined by methylation followed by acid hydrolysis which gave 3-hydroxy-4',5-dimethoxystilbene.

The hydroxystilbenes and flavonoids may share a common purpose as precursors of the condensed tannins; according to present theories they also have a close metabolic relationship since for both classes of compound one aromatic ring is derived via the acetate pathway and the other from shikimic acid[142]. This biological relationship was first postulated by Erdtmann[143] and supported by later workers[144] who indicated similarities of structure between the hydroxy stilbenes and flavonoid compounds from a particular plant source. Thus many pine species yield pinosylvin (90) and its monomethyl ether, the flavone chrysin (91) and the 2,3-dihydroflavone pinocembrin (92) all of which have a characteristic unsubstituted phenyl ring. It is probable that additional interesting biogenetic relationships such as these will be revealed during more detailed surveys of woody species and thus provide further information regarding the pathways of biosynthesis of these compounds.

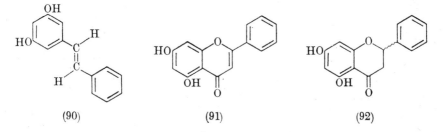

(90)   (91)   (92)

REFERENCES

1. Hathway, D. E., "Wood Extractives", (W. E. Hillis, ed.), Academic Press, New York (1962).
2. Official methods of analysis, *J. Soc. Leath. Trades Chem.*, (1957).
3. Grassmann, W. E., and Zeschitz, E., *Leder*, **3**, 241 (1952).
4. White, T., "Chemistry of Vegetable Tannins", Soc. Leath. Trades Chem., Croydon 1 (1956).
5. Freudenberg, K., "Die Chemie der Naturlichen Gerbstoffe", Springer-Verlag, Berlin (1920).
6. Kostanecki, st., and Lampe, V., *Ber. dt. chem. Ges.*, **39**, 4007 (1906).
7. Runge, F. F., *Wiss. Phytochem.*, 245 (1821).
8. Freudenberg, K., *Ber. dt. chem. Ges.*, **53**, 1416 (1920).
9. Freudenberg, K., and Purrmann, L., *Justus Liebigs Annln Chem.*, **437**, 274 (1924).
10. Bate-Smith, E. C., and Swain, T., *Chemy Ind.*, 377 (1953).
11. Bate-Smith, E. C., and Swain, T., "Chemistry of the Vegetable Tannins", Soc. Leath. Trades Chem., Croydon (1956).
12. Rosenheim, O., *Biochem. J.*, **14**, 178 (1920).
13. Robinson, G. M., and Robinson, R., *Biochem. J.*, **25**, 1687 (1931); *Biochem. J.*, **27**, 206 (1933).
14. Bate-Smith, E. C., *Biochem. J.*, **58**, 122 (1954).
15. Bate-Smith, E. C., and Lerner, N. H., *Biochem. J.*, **58**, 126 (1954).
16. Roux, D. G., *Nature, Lond.*, **181**, 1454 (1958).
17. Hathway, D. E., and Seakins, J. W., *Biochem. J.*, **72**, 369 (1959).
18. Grassmann, W. E., and Endres, H., *Leder*, **10**, 237 (1959).
19. King, H. G. C., and White, T., *Proc. chem. Soc.*, 341 (1957).

20. Cunningham, J., Haslam, E., and Haworth, R. D., *J. chem. Soc.*, 2875 (1963).
21. Hillis, W. E., "Wood Extractives", (Hillis, W. E., ed.), Academic Press, London (1962).
22. Harborne, J. B., *J. Chromat.*, **2**, 581 (1959).
23. Seikel, M. K., "The Chemistry of Flavonoid Compounds" (T. A. Geissmann, ed.), Pergamon Press, London (1962).
24. Roux, D. G., Maihs, A. E., and Paulus, E., *J. Chromat.*, **5**, 9 (1961).
25. Cahn, R. S., Ingold, C. K., and Prelog, V., *Experentia*, **12**, 81 (1956).
26. Birch, A. J., Clark-Lewis, J. W., and Robertson, A. V., *J. chem. Soc.*, 3586 (1957).
27. Roberts, E. A. H., and Wood, D. J., *Biochem. J.*, **53**, 332 (1953).
28. Mehta, P. P., and Whalley, W. B., *J. chem. Soc.*, 5327 (1963).
29. Philbin, E. M., and Wheeler, T. S., *J. org. Chem.*, **27**, 4114 (1962).
30. Philbin, E. M., and Wheeler, T. S., *Proc. chem. Soc.*, 167 (1958).
31. Roux, D. G., and Maihs, A. E., *J. Chromat.*, **4**, 65 (1960).
32. Kirby, K. S., Knowles, E., and White, T., *J. Soc. Leath. Trades Chem.*, **37**, 283 (1953).
33. Partridge, S. M., *Biochem. J.*, **42**, 238 (1948).
34. Lindstedt, G., *Acta chem. scand.*, **4**, 448 (1950).
35. Roux, D. G., *Nature, Lond.*, **180**, 973 (1957).
36. Pigman, W., Anderson, E., Fischer, R., Buchanan, M. A., and Browning, B. L., *Tech. Pap. Addr. tech. Ass. Pulp Pap. Ind.*, **36**, 4 (1953).
37. Harborne, J. B., *Fortschr. Chem. org. NatStoffe*, **20**, 165 (1962).
38. Randerath, K., "Thin-layer Chromatography", Academic Press, New York (1963).
39. Pastuka, G., *Z. analyt. Chem.*, **179**, 355 (1961).
40. Birkofer, L., Kaiser, C., Meyer-Stoll, H. A., and Suppan, F., *Z. Naturf.*, **17B**, 352 (1962).
41. Stahl, E., and Schorn, P. J., *Hoppe-Seyler's Z. Physiol. Chem.*, **325**, 263 (1961).
42. Hess, D., and Meyer, C., *Z. Naturf.*, **17B**, 853 (1962).
43. Narasimachari, N., and von Rudloff, E., *Can. J. Chem.*, **40**, 1118 (1962).
44. Pridham, J., *J. Chromat.*, **2**, 605 (1959).
45. Sato, A., and von Rudloff, E., *Can. J. Chem.*, **41**, 2165 (1963); *Can. J. Chem.*, **42**, 635 (1964).
46. Seshadri, T. R., "Chemistry of Flavonoid Compounds", (T. A. Geissmann, ed.), Pergamon Press, London (1962).
47. Schmidt, O.Th., "Moderne methoden der Pflanzenanalyse", Vol. III, p. 517 (K. Paech and M. V. Tracey, eds.), Springer Verlag, Berlin (1956).
48. Freudenberg, K., Bohme, L., and Purrmann, L., *Ber. dt. chem. Ges.*, **55**, 1734 (1922).
49. Mayer, W., and Merger, F., *Justus Liebigs Annln Chem.*, **644**, 65 (1961).
50. Weinges, K., *Justus Liebigs Annln Chem.*, **615**, 205 (1958).
51. Hörhammer, L., and Wagner, H., *Arch. Pharm., Berl.*, **289**, 316, 532 (1956).
52. Roux, D. G., and Paulus, E., *Biochem. J.*, **82**, 324 (1962).
53. Roux, D. G., and Paulus E., *Biochem. J.*, **78**, 785 (1961).
54. Roux, D. G., and Maihs, A. E., *Biochem. J.*, **74**, 44 (1960).
55. Roux, D. G., Maihs, A. E., and Paulus, E., *Biochem. J.*, **78**, 834 (1961).
56. Bradfield, A. E., and Penney, M., *J. chem. Soc.*, 2249 (1948).

57. Roberts, E. A. H., and Myers, M., *J. Sci. Fd Agric.*, **11**, 153 (1960).
58. Ice, C. H., and Wender, S. H., *J. Amer. chem. Soc.*, **50**, 75 (1953).
59. Hörhammer, L., Wagner, H., and Leeb, W., *Naturwissenschaften*, **44**, 1 (1957); *Arch. Pharm., Berl.*, **293/65**, 264 (1960).
60. Hathway, D. E., *Biochem. J.*, **70**, 34 (1958).
61. Mayer, W., "Chemistry of Vegetable Tannins", Soc. Leath. Trades Chem., Croydon (1956).
62. "Technique of Organic chemistry XI" (A. Weissberger, ed.), Interscience, New York and London (1963).
63. Geissmann, T. A., "Moderne Methoden der Pflanzenanalyse III", p. 450 (K. Paech and M. V. Tracey, eds.), Springer-Verlag, Berlin (1955).
64. Jurd, L., "The Chemistry of Flavonoid Compounds", (T. A. Geissmann, ed.), Pergamon Press, London (1962).
65. Hergert, H. L., and Kurth, E. F., *J. Am. chem. Soc.*, **75**, 622 (1953).
66. Batterham, T. J., and Highet, R. J., *Aust. J. Chem.*, **17**, 428 (1964).
67. Dyke, S. F., Ollis, W. D., and Sainsbury, H. F., *Proc. chem. Soc.*, 179 (1963).
68. Crombie, L., and Whiting, D. A., *Tetrahedron Lett.*, **18**, 801 (1962); *J. chem. Soc.*, 1569 (1963).
69. Nair, P. M., Rao, A. V. R., and Venkataraman, T., *Tetrahedron Lett.*, **2**, 125 (1964).
70. Finch, N., and Ollis, W. D., *Proc. chem. Soc.*, **176**, 177 (1960).
71. Burrows, B. F., Ollis, W. D., and Jackman, L. M., *Proc. chem. Soc.*, 177 (1960).
72. Eyton, W. B., Ollis, W. D., Sutherland, I. O., Jackman, L. M., Gottlieb, O. R., and Magalhaes, H. T., *Proc. chem. Soc.*, 301 (1962).
73. Schwarz, J. S. P., Cohen, A. I., Ollis, W. D., Kaczka, E. A., and Jackman, L. M., *Tetrahedron*, **20**, 1317 (1964).
74. Dyke, S. F., Ollis, W. D., Sainsbury, H. F., and Schwarz, J. S. P., *Tetrahedron*, **20**, 1331 (1964).
75. Clark-Lewis, J. W., and Jackman, L. M., *Proc. chem. Soc.*, 165 (1961).
76. Clark-Lewis, J. W., Jackman, L. M., and Williams, L. R., *J. chem. Soc.* 3859 (1962).
77. Corey, E. J., Philbin, E. M., and Wheeler, T. S., *Tetrahedron Lett.*, **13**, 429 (1961).
78. Clark-Lewis, J. W., Jackman, L. M., and Spotswood, T. M., *Austl. J. Chem.*, **17**, 632 (1964).
79. Crombie, L., and Lown, J. W., *J. chem. Soc.*, 775 (1962).
80. Reed, R. I., and Wilson, J. M., *J. chem. Soc.*, 5949 (1963).
81. Shaw, B. L., and Simpson, T. H., *J. chem. Soc.*, 655 (1955).
82. Looker, J. H., and Hanneman, W. H., *J. org. Chem.*, **27**, 381 (1962).
83. Sutherland, I. O., and Ollis, W. D., "Recent Developments in the Chemistry of Natural Phenolic Compounds", (W. D. Ollis, ed.), Pergamon Press, London (1961).
84. Geissmann, T. A., and Harborne, J. B., *J. Am. chem. Soc.*, **78**, 832 (1956).
85. Waiss, A. C., Lundin, R. E., and Stern, D. J., *Tetrahedron Lett.*, **10**, 513 (1964).
86. King, F. E., Clark-Lewis, J. W., and Forbes, W. F., *J. chem. Soc.*, 2948 (1955).
87. Freudenberg, K., Carrara, G., and Cohn, E., *Justus Liebigs Annln. Chem.*, **446**, 87 (1925).

88. Weinges, K., *Proc. chem. Soc.*, 138 (1964); *Justus Liebigs Annln Chem.*, **681**, 154 (1965); Whalley, W. B., "The Chemistry of Flavonoid Compounds", (T. A. Geissmann, ed.), Pergamon Press, London (1962); Clark-Lewis, J. W., and Ramsay, G. C., *Proc. chem. Soc.*, 359 (1960).

89. Aniruhdan, C. A., Mathieson, D. W., and Whalley, W. B., *Proc. chem. Soc.*, 84 (1964).

90. Pratt, D. D., Robinson, R., and Williams, P. N., *J. chem. Soc.*, 205 (1924).

91. Freudenberg, K., Fikentscher, H., Harder, M., and Schmidt, O.Th., *Justus Liebigs Annln Chem.*, **444**, 135 (1925).

92. Clark-Lewis, J. W., and Korytnyk, W. W., *J. chem. Soc.*, 2367 (1958).

93. Cummins, B., Donelly, D. M. X., Eades, J. C., Fletcher, H., O'Cinncide, F., Philbin, E. M., Swirski, J., Wheeler, T. S., and Wilson, R. K., *Tetrahedron*, **19**, 499 (1963).

94. Gowan, J. E., Hayden, P. M., and Wheeler, T. S., *J. chem. Soc.*, 862 (1955).

95. Philbin, E. M., and Wheeler, T. S., "Chemistry of Natural and Synthetic Colouring Matters", (T. S. Gore, B. S. Joshi, S. V. Sunthankar and B. D. Tilak, eds.), Academic Press, London (1962).

96. Weinges, K., *Justus Liebigs Annln Chem.*, **615**, 203 (1958); *Justus Liebigs Annln Chem.*, **627**, 229 (1959).

97. Prelog, V., *Helv. chim. Acta*, **36**, 308, 305 (1953).

98. Hardegger, E., Gempeler, H., and Züst, A., *Helv. chim. Acta*, **40**, 1819 (1957).

99. Hardegger, E., Züst, A., and Lohse, F., *Helv. chim. Acta*, **43**, 1274 (1950).

100. Roux, D. G., and Paulus, E., *Biochem. J.*, **78**, 120 (1961).

101. Hillis, W. E., and Carle, A., *Austl. J. Chem.*, **13**, 390 (1960).

102. King, F. E., and Bottomley, W., *J. chem. Soc.*, 1399 (1954).

103. King, F. E., and Clark-Lewis, J. W., *J. chem. Soc.*, 3384 (1955).

104. Clark-Lewis, J. W., and Mortimer, P. I., *J. chem. Soc.*, 4106 (1950).

105. Bottomley, W., *Chemy Ind.*, 170 (1956).

106. Bokadia, M. M., Brown, B. R., and Cummings, W., *J. chem. Soc.*, 3308 (1960).

107. Clark-Lewis, J. W., Katekar, G. F., and Mortimer, P. I., *J. chem. Soc.*, 499 (1961).

108. Freudenberg, K., and Weinges, K., *Chemy Ind.*, 486 (1959).

109. Clark-Lewis, J. W., and Katekar, G. F., *Proc. chem. Soc.*, 345 (1960).

110. Bokadia, M. M., Brown, B. R., Kolker, P. L., Love, C. W., Newbould, J. A., Somerfield, G. A., and Wood, P. M., *J. chem. Soc.*, 4663 (1961).

111. Bognar, R., Rakosi, M., Fletcher, H., Philbin, E. M., and Wheeler, T. S., *Tetrahedron Lett.*, **19**, 4 (1959).

112. Clark-Lewis, J. W., and Williams, L. R., *Austl. J. Chem.*, **16**, 869 (1963).

113. Roux, D. G., and Drewes, S. E., *Biochem. J.*, **90**, 343 (1964).

114. Joshi, C. G., and Kulkarni, A. B., *Chemy Ind.*, 1421 (1954).

115. Kashikar, M. D., and Kulkarni, A. B., *Chemy Ind.*, 1084 (1958).

116. Huitric, A. C., Carr, J. B., Trager, W. F., and Nist, B. J., *Tetrahedron*, **19**, 2145 (1963).

117. Karplus, M., *J. chem. Phys.*, **30**, 11 (1959).

118. Lemieux, R. U., Kullnig, R. K., and Moir, R. Y., *J. Am. chem. Soc.*, **80**, 2237 (1958).

119. Clark-Lewis, J. W., and Katekar, G. F., *J. chem. Soc.*, 4502 (1962).

120. Keppler, H. H., *J. chem. Soc.*, 2721 (1957).

121. Clark-Lewis, J. W., and Mitsuno, M., *J. chem. Soc.*, 1724 (1958).
122. Laumas, K. R., and Seshadri, T. R., *J. scient. ind. Res.*, **17B**, 44. (1958).
123. Ganguly, A. K., and Seshadri, T. R., *Tetrahedron*, **6**, 21 (1959).
124. Ganguly, A. K., and Seshadri, T. R., *J. scient. ind. Res.*, **17B**, 168 (1958).
125. Ganguly, A. K., Seshadri, T. R., and Subramanian, P., *Tetrahedron*, **3**, 225 (1958).
126. Roux, D. G., and de Bruyn, G. C., *Biochem. J.*, **87**, 439 (1963).
127. Chan, W. R., Forsyth, W. G. C., and Hassall, C. H., *J. chem. Soc.*, 3174 (1958).
128. Freudenberg, K., *Experentia*, **16**, 101 (1960).
129. Chakravartz, G., and Seshadri, T. R., *Tetrahedron Lett.*, 787 (1962).
130. Nair, P. M., and Venkataraman, K., *Tetrahedron Lett.*, 317 (1963).
131. Freudenberg, K., and Weinges, K., *Tetrahedron Lett.*, **8**, 267 (1963).
132. Forsyth, W. G. C., and Roberts, J. B., *Biochem. J.*, **74**, 374 (1960).
133. Erdtman, H., *Justus Liebigs Annln Chem.*, **539**, 116 (1939).
134. Späth, E., and Schlager, J., *Ber. dt. chem. Ges.*, **73**, 881 (1940).
135. Takaska, M., *J. chem. Soc. Japan*, **61**, 30 (1940).
136. Barnes, R. A., and Gerber, N. N., *J. Am. chem. Soc.*, **77**, 3259 (1955).
137. Mongolsuk, S., Robertson, A., and Towers, R., *J. chem. Soc.*, 2231 (1957).
138. King, F. E., King, T. J., Godson, D. H., and Manning, L. C., *J. chem. Soc.*, 4477 (1956).
139. Kawamura, S., *J. pharm. Soc. Japan*, **58**, 83 (1938).
140. Hillis, W. E., and Carle, A., *Biochem. J.*, **82**, 435 (1962).
141. Takaoka, S., *J. chem. Soc. Japan*, **61**, 30, 1067 (1940).
142. Birch, A. J., *Fortschr. Chem. org. NatStoffe*, **14**, 189 (1957).
143. Erdtman, H., "Progress in Organic Chemistry—I", p. 27 (J. W. Cook, ed.), Butterworths, London (1952).
144. Lindstedt, G., and Misiorny, A., *Acta. chem. scand.*, **5**, 121 (1951).

CHAPTER 3

Condensed Tannins – Structure

I. Chemical Degradation

Very little is known about the structure of the condensed tannins in spite of a great deal of chemical investigation over a number of years. A survey of the literature shows that much of the early work on these polymeric materials aimed at too great a simplification of their chemistry, many workers attempting to express the properties of each tannin within one structural formula. The problems presented to the organic chemist by the condensed tannins are analogous in many respects to those in the determination of the structure of the other complex natural phenolic polymers such as humic acid and lignin. It seems probable moreover that the solution to the structure of the condensed tannins will be expressed in terms similar to those used by Freudenberg[1] to describe the structure of the lignins. The present state of knowledge of the constitution of lignin rests primarily on the advances made towards elucidating its mode of biogenesis, which it is assumed proceeds by an enzymatically controlled oxidative polymerization of the monomolecular C_6–C_3 precursors (1, 2, 3) of which coniferyl alcohol (2) is thought to be the principal participant. Support for this hypothesis comes from the observation[2] that a phenol oxidase, laccase, catalyses the oxidative polymerization of (2) to a product closely resembling in chemical and physical properties natural conifer lignin. The isolation and characterization of simple dimeric products from this *in vitro* enzymic dehydrogenation of coniferyl alcohol (2) and a consideration of their possible modes of generation from the monomer has permitted Freudenberg[1] to elaborate a chemically plausible pathway for the formation of lignin from the cinnamyl alcohols (1, 2, 3). The picture of the structure of lignin which has emerged is that of a completely different class of natural macromolecule from say the proteins and nucleic acids where the major linkages between monomer units are essentially the same. Polymerization of the monomer (2) to give lignin proceeds in a less rigid manner with the formation of a variety of inter-monomer linkages of both the C—O and C—C types. Limitations on the ultimate structure of the lignins derived by this process are placed by the need to conform to the results of chemical degradations on the natural lignin; these reveal the number and type of functional groups (e.g. —OH, —OMe, —C—O—C—) in the polymer and in the structures of many of the simple fragments (obtained for instance by oxidation) the types of linkage between aromatic nuclei.

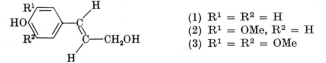

(1) $R^1 = R^2 = H$
(2) $R^1 = OMe, R^2 = H$
(3) $R^1 = R^2 = OMe$

Research on the structure of the condensed tannins is proceeding along parallel although less well defined lines and the progress made so far falls well short of that made with the structure of lignin. Thus although considerable advances have been made towards the identification of the natural precursors of these tannins such as the flavan-3-ols and flavan-3,4-diols and of the various ways in which these compounds may polymerize to form tannin-like materials, little significant improvements in our knowledge of the structural patterns present in the natural tannins themselves have resulted in the last 30 years. Undoubtedly the major difficulty encountered in these investigations has been the formulation of methods for the isolation and analysis of the condensed tannins in what may be loosely termed a "homogeneous state" and in particular uncontaminated with monomeric phenols of a flavonoid nature. Some progress in this direction has been made, notably by Roux[3] who has isolated a series of optically active tannins from the heartwood of *Acacia mollisima* (syn *A. mearnsii*) by employing a combination of counter-current distribution and thick paper chromatographic separations. The tannins, which were characterized by their paper chromatographic patterns, correspond in molecular weight to the tri-, penta- and decameric forms of the flavan-3,4-diol (+)-mollisacacidin (4) which accompanies them in the heartwood. Subsequent chemical degradation of the tannins clearly implicated the associated flavan-3,4-diol (4) in their formation and is one of the more satisfactory pieces of evidence available for regarding the flavan-3,4-diols as one of the ultimate precursors of these tannins. All the oligomers on acid treatment gave fisetinidin chloride (5) demonstrating that the mode of condensation of the flavan-3,4-diol is such that one at least of the monomer units remains potentially available to give the corresponding anthocyanidin,

Mild acid hydrolysis of the trimeric tannin gave a substance which was identified paper chromatographically as the flavan-3,4-diol (4) and these observations are most satisfactorily interpreted in terms of an acid labile ether linkage between some of the flavan-3,4-diol units in the tannin. The type of linkage envisaged may be similar to those which have been suggested in the structures of the two closely related pro-anthocyanidins isolated by Freudenberg and Weinges[4] and Forsyth and Roberts[5] from *Crataegus oxyacantha* and cacao respectively. Both these compounds have properties which simulate to some extent the behaviour of the flavan-3,4-diol oligomers from *Acacia mollisima* and

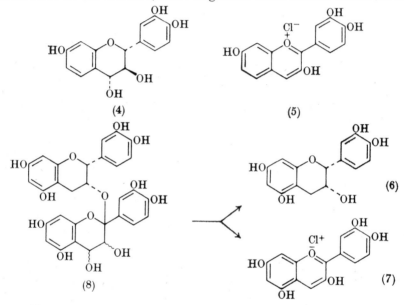

on acid treatment gave (−)-epicatechin (6) and cyanidin (7). Structure (8) which has been suggested[4] for the proanthocyanidin from *Crataegus oxyacantha* is illustrative of one of the possible ways in which flavan-3-ols and flavan-3,4-diols may condense together to form dimeric products and may typify one of the ways in which these compounds polymerize in Nature to form tannin-like materials.

The results of Roux on the tannins and flavonoids of *Acacia mollisima* heartwood also bring out in another way the intimate connection which the flavan-3,4-diols have with condensed tannin formation. Thus it was observed[6] in a semi-quantitative fashion that the concentration of the flavan-3,4-diol (+)-mollisacacidin (4) was highest in the peripheral heartwood and decreased in concentration in passing towards the centre of the heartwood. Parallel with this change the concentration of the condensed tannin and the related flavonol and 2,3-dihydroflavonol increased with increasing age of the heartwood. Roux has interpreted these observations[6] as defining a possible biogenetic sequence in which the flavan-3,4-diol is the first formed flavonoid which then undergoes a progressive transformation to chemically more stable flavonoid compounds and to the condensed tannins associated with this wood. Similar schemes have been suggested for tannin formation and the biogenesis of flavonoids in the woods of *Schinopsis* species[7] and *Robinia pseudacacia*[8].

Direct evidence for the participation of flavan-3-ols in condensed tannin production has been obtained by Freudenberg and Weinges[9] who isolated a dimeric form (9) of (+)-catechin by counter-current distribution of *Uncaria gambir* extract. The isolation of (9) from a natural source lends additional support to Freudenberg's views[10], which have been deduced from *in vitro* experiments, regarding the mechanism of catechin polymerization leading to the production of tannins, since structure (9) had previously been assigned to the dimeric product resulting from acid catalysed condensation of (+)-catechin. Several detailed studies concerning the mechanism of the transformation *in vitro* of flavan-3-ols (and flavan-3,4-diols) to tannins have been carried out but, apart from the isolation of (9) from a natural source, evidence which will link the structural patterns present in the tannins to one or more of the possible modes of condensation or polymerization of the precursors is still largely missing. Until thorough investigations along these lines have been pursued the picture of the structure of the condensed tannins will remain incomplete.

II. The Condensation–Polymerization Reactions of Flavan-3-ols and Flavan-3,4-diols

A number of alternative routes for the conversion of the flavan progenitors to the condensed tannins have been considered, varying from the acid catalysed mechanism favoured by Freudenberg[10] to the oxidative polymerization discussed by Hathway and Seakins[11, 12]. The suggestions of a simulated acid catalysed mechanism for the formation of tannins from (+)-catechin put forward by Freudenberg have been discounted by several workers[13] on the grounds that the conditions of low pH required would not be encountered in any living or dead plant tissue. However, since the flavan-3-ol → tannin conversion undoubtedly takes place at a very much slower rate in plants than in the laboratory, the above arguments against the Freudenberg mechanism do not appear to be substantial ones. In addition, with many woods and extracts it is not clear whether part of the tannin which is recovered is formed during the extraction and concentration and consideration of the Freudenberg mode of condensation is certainly necessary in so far as these processes are concerned. Freudenberg[14] supports the view that the formation of the natural condensed tannin is a spontaneous post-mortal process which encompasses many years in the non-living cells of the woody plant; the polymerization may, however, be enzymically controlled in living tissues and this point still appears open to question. Nevertheless both the acid catalysed and

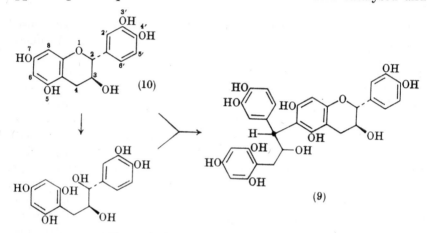

FIG. 1. Self condensation of (+)-catechin[15].

oxidative processes of flavan polymerization are considered here in so far as both give rise to amorphous tannin-like materials and are therefore of importance in any general consideration of the reactions which may be involved in the biogenesis or formation of the condensed tannins.

The acid catalysed condensation of (+)-catechin (10) and related hydroxyflavans has been studied extensively by Freudenberg, Mayer and Brown and their colleagues. A mechanism for the self condensation of (+)-catechin was first suggested in 1934 by Freudenberg and Maitland[15] (Fig. 1) and in subsequent work Freudenberg and Alonso[16] were able to isolate from this reaction a dicatechin (9) as its undeca-acetate and hence confirm in outline the proposed pathway. Freudenberg and Weinges[17, 18] have also carried out a more general study on the self-condensation of flavan derivatives and have shown that the minimum structural features necessary are a *para*-hydroxy (or methoxy) benzyl ether group (i.e. —OH or —OMe group in 4' position of flavan nucleus) and a resorcinol orientation of hydroxyl, or potential hydroxyl, groups in ring A (i.e. an —OH group in the 7 position). Thus they were able to show[18] that the simplest flavan which would undergo self-condensation was 4',7-dihydroxyflavan (11); in this case polymerization occurred in 3–5 min in boiling acid or 3 days at pH 5 and room temperature. Conversely they observed that substitution of the 6 or 8 position of a 7-hydroxyflavan derivative by methyl or chlorine inhibited the condensation and these facts are rationalized in the mechanism for the self-polymerization of (+)-catechin shown in Fig. 2. It is suggested[10] that (+)-catechin (10) acts at C_2 as an electrophile (resonance stabilization of a carbonium ion at this centre being promoted by the 4' hydroxyl group (10a)) and is thus able to substitute a further molecule of (+)-catechin at C_6 or C_8 ortho to a hydroxyl group to give the dimer (9).

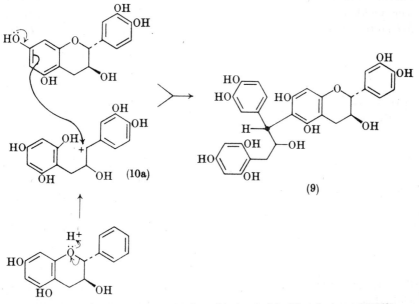

FIG. 2. Acid catalysed polymerization of (+)-catechin (Freudenberg, 1958[10]).

Support for this mechanism also comes from the work of Brown and Cummings[19] who condensed 4'-methoxyflavan and 7-hydroxyflavan to give the dimer (12) isolated as its diacetate. In this model reaction the electrophilic entity provided by the 4'-methoxyflavan molecule must substitute ortho to the 7-hydroxy group of the 7-hydroxy flavan in a manner entirely analogous to that described by Freudenberg[10] for (+)-catechin (Fig. 2).

An alternative mechanism for the acid catalysed self-condensation of (+)-catechin (10) proposed by Mayer and Merger[20] has been revised in the light of a more detailed investigation[21–24] of the model reaction between (+)-catechin and phloroglucinol[25] upon which it was based. Reaction of (+)-catechin and phloroglucinol gave a compound $C_{21}H_{18}O_8$ corresponding to the loss of one molecule of water between the two reacting species. The same compound was also formed by condensation of (−)-epicatechin and phloroglucinol and gave a hepta-acetate and with diazomethane a heptamethyl ether whose infra-red spectrum showed the absence of aliphatic hydroxyl groups. Three possible structures (13, 14, 15; $R^1 = R^2 = H$) were envisaged for this compound and formation of (14 or 15; $R^1 = R^2 = H$) it was suggested would involve a characteristic Wagner–Meerwein rearrangement of the ether oxygen or 2-aryl group of (+)-catechin during the course of the reaction. Condensation of ethyl phloroglucinol with (+)-catechin however yielded[22] two isomeric products which can therefore only be satisfactorily interpreted in terms of structure (13; $R^1 = C_2H_5$,

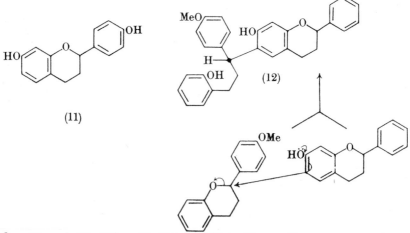

(11)   (12)

$R^2 = H$ and 13; $R^1 = H$, $R^2 = C_2H_5$). Hence the structure of the original condensation product of (+)-catechin and phloroglucinol must be (13; $R^1 = R^2 = H$) and this was confirmed by Mayer and Lemke[21] in a synthesis of the methylated dehydrogenation product (16) (Fig. 3). Further corroborative evidence for the structure (16) has been obtained by Freudenberg and Weinges and their colleagues[23, 24] who prepared (16) by acid catalysed condensation of phloroglucinol with (17) followed by methylation of the product. This reaction in order to give (16) must involve both fission of an ether linkage and re-etherification and contrasts with the parallel reaction with resorcinol which proceeds in the expected fashion to give, after methylation, (18). The structures (16) and (18) were both fully supported by proton magnetic resonance measurements.

Condensation of resorcinol with (+)-catechin or its methyl ether

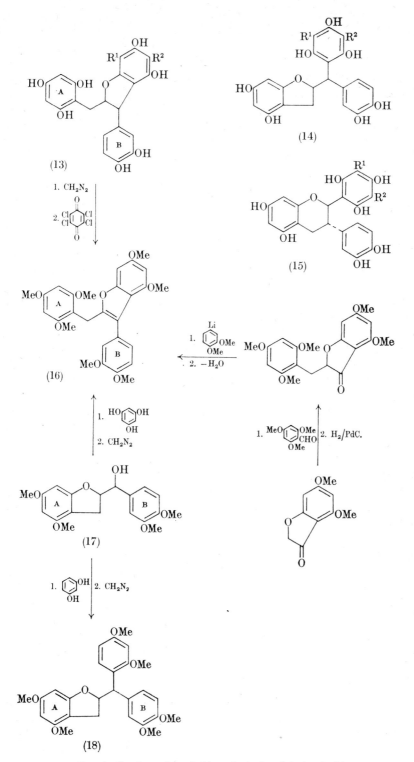

FIG. 3. Condensation of phloroglucinol and (+)-catechin.

gave after methylation a product from which resorcinol dimethyl ether
was obtained by alkali fusion and whose proton magnetic resonance
spectrum differed appreciably from that of the analogous phloroglucinol
condensate. It is probable therefore that in its condensation with (+)
catechin (as with (17)) resorcinol and its derivatives give products in
which both the original resorcinol hydroxyl groups remain free.†

Mayer and Merger[20] in their study of (+)-catechin self-condensation
treated (+)-catechin at 90° and pH 4 and obtained from the reaction
(+)-epicatechin (by epimerization at C_2 of the original (+)-catechin)
and two epimeric dicatechin derivatives, one of which was obtained in
a crystalline form. The structures (19 or 20) assigned to the dimers by
Mayer and Merger[20] were based on the observation that they resulted
from loss of one molecule of water from two (+)-catechin molecules
and on their original interpretation of the model reaction between
(+)-catechin and phloroglucinol which favoured a Wagner–Meerwein
change involving either the ether oxygen or the 2-aryl group of (+)-
catechin upon attack of the nucleophile at C_2. The dimeric products
were assumed to be epimers at the centres (*) shown (19 or 20). However,
Freudenberg[9] later showed that the original dicatechin (9) obtained by
treatment of (+)-catechin at 25° and pH 2 was transformed on warming
in acid at 90° to the crystalline dicatechin obtained by Mayer and
Merger and analysis showed the reaction to be a simple dehydration.
Freudenberg and Weinges in view of this evidence and the reinter-
pretation of the model reaction therefore favoured structure (19) for
the crystalline dicatechin and suggested[9] that it was formed from the
dimer (9) by the pathway shown in Fig. 4. The basis for this particular
mode of dehydration was the evidence obtained from the model reaction
between (+)-catechin and resorcinol[23, 24] which showed that in the
final product none of the hydroxyl groups of the original resorcinol
molecule were involved in cyclic ether linkages. Assuming ring A of the
second (+)-catechin molecule to act as a resorcinol derivative by analogy none of its hydroxyl groups should participate in ether linkages in
the final dehydration product (19). Compounds (9) and (19) obviously
represent the first stages in the self-condensation of (+)-catechin but
since both still retain centres of potential electrophilic and nucleophilic
character, further steps in the progressive condensation of (+)-catechin
to yield the amorphous tannin could occur in ways very similar to
those demonstrated for the initial steps of the condensation.

Hydroxyflavan-3,4-diols undergo self-condensation more readily
under the influence of acids than the corresponding flavan-3-ols and

† A recent publication (K. Weinges and F. Toribio, *Leibigs Annalen*, **681**, 163 (1965))
confirms this view.

this has led to the supposition[10, 25] that the secondary hydroxyl group in the four position of the flavan nucleus participates directly in the reaction. The immediate environment of this hydroxyl group gives it the properties of an *ortho*-alkoxybenzyl alcohol which under acidic conditions would therefore be expected to readily form the corresponding benzyl carbonium ion. Kenyon and his colleagues[26-30] investigated the general reactions of benzyl alcohols of this type and

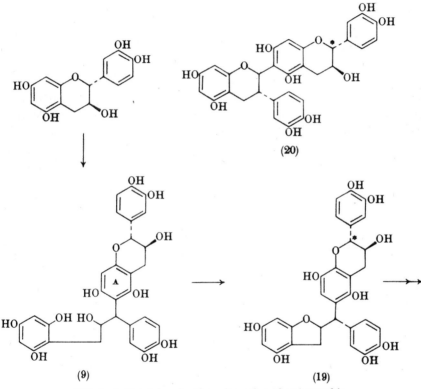

Fig. 4. Acid catalysed self-condensation of (+)-catechin.

showed[29] that compounds such as (21), where the cumulative mesomeric effect of the methoxyl groups results in an enhanced resonance stabilization of the benzyl carbonium ion, react in acidic media readily and exclusively by alkyl-oxygen fission. Thus with alcohols and thiols in hot chloroform (21) gave the corresponding ethers and with benzoyl-acetone containing a trace of sulphuric acid the *C*-alkylated derivative (22). Condensation was similarly readily effected with 2,4,6-trimethoxy- and 2,4,6-triethoxy benzene to give (23; R = OMe and R = OEt respectively). The acetate of (21) reacted similarly and with phenols gave the corresponding *O*-aralkylated derivatives. In an extension of these observations Brown, Cummings and Newbould[31] studied the model condensation of flavan-4-ol (24) with various phenols and their results follow the pattern anticipated from the work of Kenyon and his

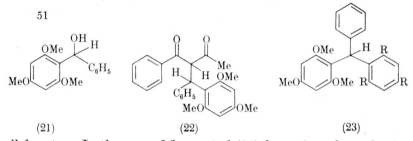

51

(21)   (22)   (23)

collaborators. In the case of flavan-4-ol (24) formation of a carbonium ion at the four position is promoted by the resonance stabilization possible with the ether oxygen and is entirely analogous to the *ortho-*

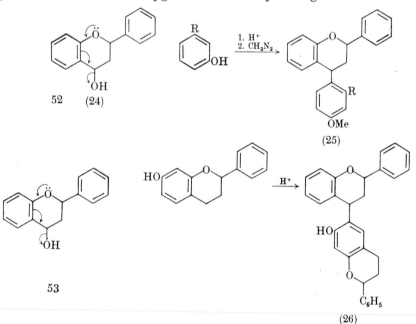

52   (24)

(25)

53

(26)

(26) alkoxybenzyl alcohols considered above. Thus with phenol or resorcinol flavan-4-ol gave, after methylation, (25; R = H or R = OMe respectively). Upon reaction with 7-hydroxyflavan a product, formulated as (26) on the basis of the known nucleophilic character of 7-hydroxyflavan at the 6 position, was obtained. The condensation of

phenol with 4′-methoxyflavan-4-ol (27) illustrates the probable diversity of pathways which may be followed in the self-condensation of hydroxy substituted flavan-3,4-diols. The molecule of 4′-methoxyflavan-4-ol combines the known electrophilic character of 4′-methoxy flavans at the 2-position and of flavan-4-ols at the 4-position, and on reaction with phenol gave after methylation a compound (28) whose structure was confirmed[31] by synthesis. Brown, Cummings and Newbould[31] also investigated the acid catalysed condensation of resorcinol and phloroglucinol with flavan-3,4-diol and isolated after methylation (29; R = H and R = OMe respectively) in which condensation as expected had occurred at the four position. Significantly no product was isolated

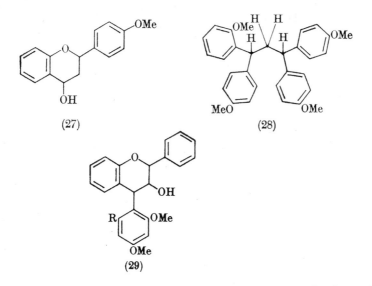

(27)   (28)

(29)

in which the 3-hydroxyl of the 3,4-diol grouping was involved in ether formation with the original resorcinol or phloroglucinol ring.

Detailed investigations on the course of acid catalysed self-condensation of a naturally occurring flavan-3,4-diol have not been reported, but it is possible on the basis of the work of Freudenberg, Mayer and Brown and their colleagues, to speculate briefly on the probable pathways of polymerization of a flavan-3,4-diol with a minimum of the 4′,7 pattern of hydroxyl substitution in the aromatic nuclei. Thus it may be surmised that hydroxyflavan-3,4-diols of this type would combine a bifunctional character similar to (+)-catechin (e.g. electrophilic at C_2 and nucleophilic at C_6 or C_8) with an additional electrophilic character at the 4 position. Freudenberg[32] has suggested on this basis that the polycondensates of flavan-3,4-diols are linked preferentially between C_4 and C_8 or C_4 and C_6 (Fig. 5, (30)). Further condensation it may be assumed could take place in a similar manner to the above and would probably also involve condensation at the C_2 position of additional flavan-3,4-diol molecules (e.g. 31). In order to

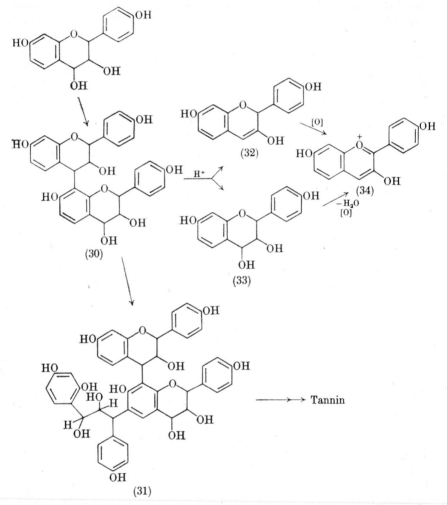

FIG. 5. Possible pathways of self-condensation of flavan-3,4-diols and acid degradation of the polymers.

accommodate the observations by Roux[3, 25] that the natural condensed tannins liberate flavan-3,4-diols or their transformation products upon acid treatment Freudenberg[32] proposed a mechanism for the degradation in acid of polymerized flavan-3,4-diols (Fig. 5). The reaction,

which entails a familiar cleavage of a substituent from its position in a
resorcinol or phloroglucinol ring, would result in the liberation of two
C_{15} fragments (32, 33) both of which could be then converted, at least
in part, to the anthocyanidin (34).

An alternative explanation of the manner of self-condensation of
flavan-3,4-diols has been put forward by Roux who has suggested[3]
that the reaction proceeds with the formation of ether linkages between
the 4 position of one flavan-3,4-diol and the 6 or 8 position of a further
unit. Although Brown and his collaborators[31] did not report any
instances of O-aralkylation by flavan-4-ols during their model experiments Clark-Lewis and Mortimer[33] showed that isomelacacidin (35;
R = H) isolated from *Acacia melanoxylon* was readily transformed by

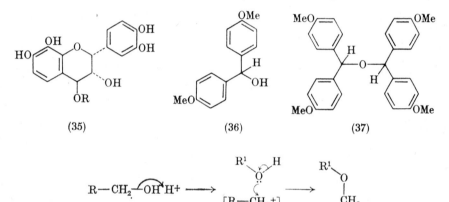

warming with ethanolic acetic acid into its O-ethyl derivative (35;
R = Et). The work of Kenyon and his colleagues has moreover
emphasized the generality of O-aralkylation reactions with activated
benzyl alcohol systems. Thus the acetate of (21) readily afforded aryl
and alkyl ethers when heated with the appropriate phenol or alcohol
and the alcohol (36) when treated[28] with half a molecular proportion of
acetic acid at 100° gave the self-condensation product (37). These and
the other O-aralkylations enumerated are most satisfactorily interpreted in terms of the reaction sequence (38 → 40) in which the
resonance stabilized benzyl carbonium ion (39) reacts with the nucleophilic hydroxylic component to give the ether (40). These model
reactions therefore provide chemically plausible pathways for the
formation (as suggested by Roux), of ether linkages between one
flavan-3,4-diol unit and a phenolic or alcoholic hydroxyl group of a

further unit (e.g. (41)) during self-condensation. Such linkages would be of the benzyl ether type and would be expected to readily cleave with aqueous acid with the resultant formation of the anthocyanidin (Fig. 6). It is perhaps surprising that the acid catalysed condensation of flavan-4-ols and (+)-catechin with other phenols have not resulted in the formation of products containing similar benzyl ether linkages but more detailed scrutiny of the products of such reactions may reveal the presence of compounds of this type.

The transformations of the hydroxy flavan-3,4-diols and hydroxy flavan-3-ols brought about by acids may be summarized as resulting from the ability of these molecules to act bifunctionally—both as an

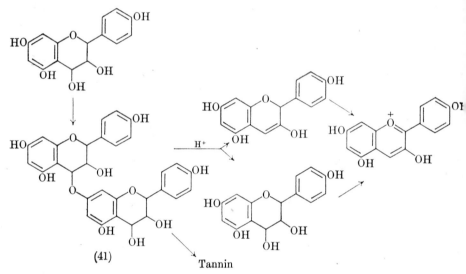

Fɪɢ. 6. Further possible modes of self-condensation of flavan-3,4-diols and degradation of the polymers.

electrophile and as a nucleophile. The model reactions which have been utilized to elucidate the precise nature of these changes have in several instances made use of the electrophile character of the flavan derivatives (at C_2 or C_4) by reaction with an external nucleophile (e.g. phenol or resorcinol). It is clear therefore that if the hypothesis of Freudenberg[14] is accepted, namely that the formation of the tannin follows a similar pathway to the acid catalysed polymerization of flavans then the presence of other natural phenols in the plant tissue will probably markedly affect the nature and composition of the resultant tannins. Thus for example spruce bark contains in addition to (+)-catechin considerable quantities of the hydroxystilbene piceatannol (42) and it is not inconceivable that the tannin of this bark is derived in part (Fig. 7) from a cross polymerization of (+)-catechin (exhibiting its electrophilic character) and piceatannol (acting as a nucleophile). The polymerization of hydroxy flavan-3,4-diols and flavan-3-ols probably represents the basic reaction for the production of condensed tannins but the complexity of the mixtures of other phenols which are present in certain tannin bearing tissues (e.g. *Schinopsis* species[7]) suggests that some at least may be incorporated into the tannin during its formation. As yet because of insufficient degradative work on the natural tannins themselves a more realistic appraisal of the merits of this modification to the basic pattern of tannin formation cannot be made.

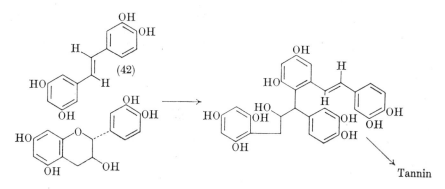

FIG. 7. Crossed condensation of (+)-catechin and piceatannol.

III. THE OXIDATIVE POLYMERIZATION OF FLAVAN-3-OLS

As an alternative to the acid-catalysed self-condensation of (+)-catechin as a model for the formation of condensed tannins Hathway and Seakins[11, 12] have made a study of the enzymic autoxidation of (+)-catechin as a possible mode of polymerization to give these complex amorphous materials. Catechin autoxidation polymer was formed when (+)-catechin was shaken with silver oxide in phosphate buffer and in a corresponding biochemical study enzymic oxidation of (+)-catechin with a variety of plant polyphenoloxidases gave a polymer with similar chemical and physical characteristics. The enzymic oxidation polymer of (+)-catechin gave several tests which indicated that it had the requisite tanning properties and it had similar absorption spectra and analytical properties to purified tannin fractions obtained from *Acacia catechu* heartwood and *Uncaria gambir* leaves. Since *Uncaria gambir* leaves contain 30–40% of (+)-catechin and *Acacia catechu* heartwood considerable quantities of the diastereoisomer (−)-epicatechin Hathway drew the reasonable conclusion that the condensed tannins from these sources are formed by an enzymatically controlled oxidation of these associated monomeric flavan derivatives. Manometric and spectroscopic evidence led Hathway and Seakins to suggest (Fig. 8) that the oxidation sequence involved the initial formation of an *ortho*-quinone (43) followed by 1–4 nucleophilic addition of a further phenolic molecule to the quinone and subsequent reoxidation to the quinonoid state. The absorption spectra of the autoxidation

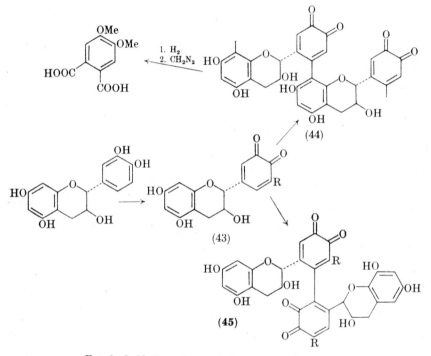

Fig. 8. Oxidative polymerization of (+)-catechin[11, 12].

polymer of (+)-catechin was found to differ considerably from those derived from the simple catechol types (e.g. 3′,4′-dihydroxyflavan and 5,7-di-*O*-methylcatechin) and Hathway and Seakins were therefore led to suggest that polymerization of (+)-catechin probably involved "head-to-tail" (44) linkages. Evidence for the presence of "head-to-tail" linkages in the polymer was obtained by Hathway[34] by degradation of catechin autoxidation polymer. Reduction followed by diazomethane methylation gave a product which on oxidation yielded small amounts of veratric and *m*-hemipinic (46) acids. The latter acid could only arise from molecules containing the "head-to-tail" type of linkage (44) and its isolation provides support for Hathway's postulates. In the case of gallocatechin it was however suggested that oxidative polymerization occurred with the predominant formation of "tail-to-tail" linkages (45; R = OH).

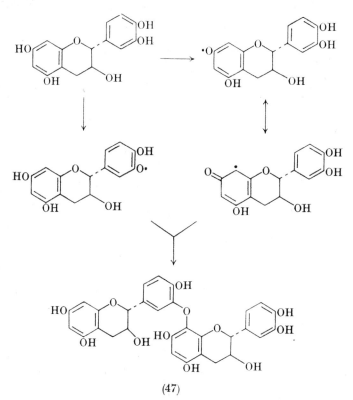

(47)

FIG. 9. Oxidation of (+)-catechin—alternative pathways.

Increasing evidence is being obtained that phenol oxidation reactions play an important role in the biosynthesis of many groups of natural products. Thus for example mechanisms involving phenol oxidation have been proposed for steps in the biosynthesis of lignin[1], thyroxine[38], various phenolic alkaloids[36] and many fungal metabolites[37, 35]. The formation of most of these products has been rationalized in many cases by assuming[36, 37, 39] that the first step is the generation of the phenolate radical from the phenol. Subsequent steps normally involve the formation of stable end products by radical pairing to give compounds containing C—C or C—O linkages dependent on the type of pairing postulated. With this more generalized picture of phenol oxidation in mind the formation of a wider variety of "inter-flavan" linkages during the enzymatically or chemically controlled oxidation of (+)-catechin would be predicted. In particular it would be reasonable to expect to find in the polymeric tannin materials some C—O linkages between flavan units formed by radical pairing as shown in Fig. 9. Similar C—O

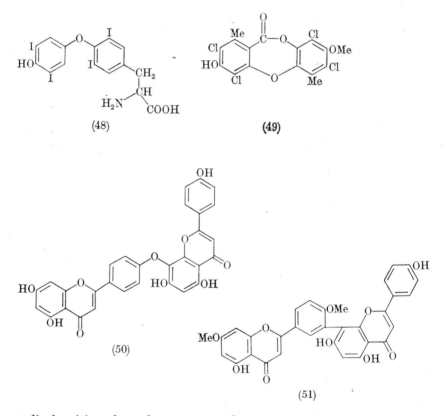

(48)   (49)

(50)

(51)

radical pairings have been suggested to account for the biogenesis of natural products such as thyroxine[38] (48) and the depsidones[40, 41] (e.g. 49) from simpler phenolic precursors. Indirect support for these predictions also comes from the finding[42] of the biflavonyl hinoki-flavone (50) with a C—O "inter-flavone" linkage in the leaves of *Chamaecyparis obtusa*. Biflavonyls in which there is a C—C linkage between the two flavone units (e.g. ginketin (51), *Ginko biloba*) are more frequent in their occurrence[43] and since the biogenesis of these compounds has been suggested as occurring by oxidative coupling of two flavone units this may indicate a preference in the oxidation of flavonoids for the formation of C—C as opposed to C—O linkages.

Further interesting possibilities of the pathways which may be followed during the oxidative polymerization of flavan derivatives are derived from the work of Roberts and his collaborators[44-50] on the chemistry of tea manufacture. The black tea of commerce is produced from the young shoots of the tea bush (*Camellia sinensis*), the total dry weight of which contains up to 30% of phenolic materials. The major phenolic constituents are (−)-epigallocatechin (52; R = H) and its 3-galloyl ester (52; R = 3,4,5-trihydroxybenzoyl) and these are shown along with the other polyphenols in the paper chromatogram

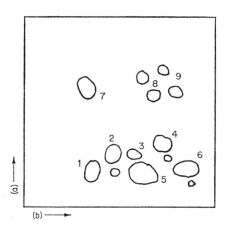

FIG. 10. Paper chromatogram of an extract of freshly plucked tea shoots. Developing solvents (a) 2% acetic acid; (b) butan-1-ol–acetic acid–water (4:1:5). Compounds identified: 1, (−)-epigallocatechin; 2, (+)-gallocatechin; 3, (−)-epicatechin; 4, (+)-catechin; 5, (−)-epigallocatechin gallate; 6, (−)-epicatechin gallate; 7, theogallin; 8, chlorogenic acids; 9, *p*-coumaroylquinic acids.

of an extract of fresh tea leaves (Fig. 10)[46]. In the conventional methods of tea manufacture the plucked shoots are allowed to wither for some hours before rolling, a process which damages the cell structure of the leaves sufficient to initiate fermentation. Finally after 3–4 hours' "fermentation" the leaves are subjected to a current of hot air (∼ 80°C) which destroys the enzyme and the product is the familiar black tea. The liquor characteristics of a tea such as colour and strength are developed during the fermentation and Roberts[47] and other workers have ascribed these to the products of enzymic oxidation of the leaf polyphenols. Thus it is noteworthy that strength and colour are most highly developed in teas derived from the terminal shoots in which the polyphenol content may be as high as 35% whereas teas of a lower grade result from the more mature leaf and stems with a lower polyphenol content (15–25%). The marked colour development which occurs during fermentation is accompanied by a decrease in concentration of (−)-epigallocatechin and its gallate and by the formation of a number of new polyphenolic compounds. A paper chromatogram[48] of a black tea extract which shows the presence of these compounds (A–F) is illustrated in Fig. 11. The fermentation process according to Roberts consists essentially of an oxidation of leaf polyphenols catalysed by oxidases. Consideration[49] of the oxidation–reduction potentials indicate

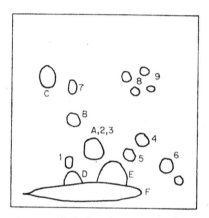

Fig. 11. Paper chromatogram of ethyl acetate extract of a black tea infusion. Solvent systems (a) and (b) and compounds 1–9 as for Fig. 10. Provisional identifications: A, digalloyl bisepigallocatechin; B, galloyl bisepigallocatechin; C, bisepigallocatchin; D, theaflavin; E, theaflavin gallate; F, thearubigins.

that (−)-epigallocatechin and its gallate are preferentially oxidized under these conditions and all the oxidation products of black tea are thought to arise from these two precursors. This hypothesis was confirmed by Roberts in studies of the enzymic oxidation of (−)-epigallocatechin and its gallate which together gave all the compounds (A–E). The thearubigins (F) represent complex polyphenolic polymers (tea tannins) and are thought to arise by further coupled oxidation of compounds (A–E) with other polyphenols and possibly amino acids and proteins in the leaf. The tentative structures put forward for compounds (A–E) and the proposed mechanisms for their formation are shown in Fig. 12. The first step it is suggested is an oxidative dimerization of (−)-epigallocatechin and or its gallate (52; R = H or 3,4,5-

trihydroxybenzoyl) similar to that suggested by Hathway for (+)-catechin polymerization. The identification of (A) as a digalloyl and (B) as a monogalloyl ester of the bisflavonol (C) (53; $R^1 = R^2 = H$) accord with their chromatographic behaviour and the fact that on enzymic hydrolysis both A and B gave C and gallic acid. Theaflavin (D) and its gallate (E) were assigned the benztropolone structures

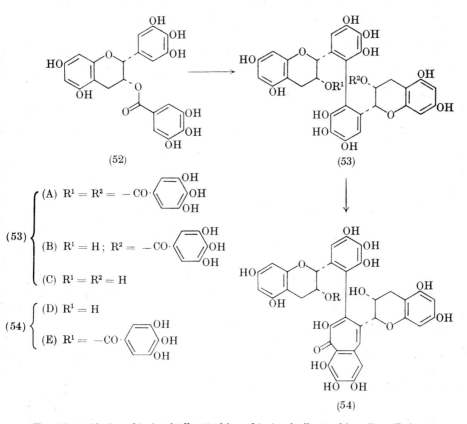

(52)

(53)

(53) {
(A) $R^1 = R^2 = -CO·$ (ring with OH, OH, OH)

(B) $R^1 = H$; $R^2 = -CO·$ (ring with OH, OH, OH)

(C) $R^1 = R^2 = H$
}

(54) {
(D) $R^1 = H$

(E) $R^1 = -CO·$ (ring with OH, OH, OH)
}

(54)

FIG. 12. Oxidation of (−)-epigallocatechin and (−)-epigallocatechin gallate (Roberts).

(54; $R^1 = H$ and $R^1 = 3,4,5$-trihydroxybenzoyl) on the basis of the close similarity in their light absorption spectra to that of purpurogallin and the observation that no theaflavins were formed when (−)-epigallocatechin was oxidized in the absence of its 3-galloyl ester. Roberts made the novel suggestion that one of the galloyl groups in the bisflavonol (A) was oxidized and coupled with one of the pyrogallol nuclei of the bisflavonol to yield the benztropolone residue of theaflavin gallate (E). An analogous transformation of (B) would give thea-flavin (D).

Studies on the mechanism of oxidation of tea leaf catechins have also been carried out by Takino and Imagawa[51] who found that a mixed oxidation of epigallocatechin and *epicatechin*, using tea oxidase or potassium ferricyanide, gave a crystalline orange pigment to which they assigned structure (55) from a consideration of its spectral characteristics. The compound displays remarkable similarities to the natural theaflavin but a full comparison has not been made, neither has the structure (55) been fully substantiated. Further evidence is clearly desirable to prove either of the novel structures (54, D) or (55) put forward for theaflavin.† Takino and Imagawa[52, 53] also performed a series of mixed oxidations between catechol derivatives and flavonoid

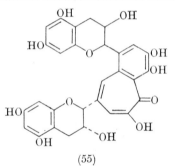

(55)

compounds with a 3,4,5-trihydroxylated B ring. The results of these model reactions suggest that the formation of benztropolones is probably a major pathway of oxidative polymerization of any flavonoid with an *o*-trihydroxylated B ring and it is clear that these ideas have a wider application than in the context of tea manufacture if oxidative pathways such as those postulated by Hathway and Seakins play a part in condensed tannin formation.

The picture presented of the condensed tannins is thus a very biased one. Considerable amounts of evidence have been accumulated which permit the elaboration of detailed pathways for the formation of the tannins by acid or oxidative catalysis from flavan type precursors and in this respect the catechin hypothesis still holds the centre of the stage. Whilst it is evident that further research is necessary to elucidate the finer points of catechin *in vitro* polymerization, it is also clear that the most significant discoveries in this field of study remain to be made by the application to the natural tannins of novel methods of breakdown and analysis of the fragments obtained. The potentialities of work in

† Recent work has confirmed structure (55) for theaflavin.

this field were recently made evident by a report[54] that condensed tannins were readily degraded with sulphurous acid to simpler products and that in their response to such treatment the natural polymers from cacao and mangrove resembled one another but differed considerably from the synthetic catechin polymers produced by the *in vitro* methods described above.

REFERENCES

1. Freudenberg, K., *J. pure appl. Chem.*, **5**, 9 (1962); *Nature, Lond.*, **183**, 1152 (1959).
2. Freudenberg, K., Harkin, J., Reichert, M., and Fukuzumi, T., *Chem. Ber.*, **91**, 581 (1958).
3. Roux, D. G., and Paulus, E., *Biochem J.*, **78**, 476 (1961).
4. Freudenberg, K., and Weinges, K., *Tetrahedron Lett.*, **8**, 267 (1961).
5. Forsyth, W. G. C., and Roberts, J. B., *Chemy Ind.*, 755 (1958).
6. Roux, D. G., and Maihs, E. A., *Biochem. J.*, **78**, 123 (1961).
7. Roux, D. G., and Paulus, E., *Biochem. J.*, **78**, 785 (1961).
8. Roux, D. G., and Paulus, E., *Biochem. J.*, **82**, 320 (1962).
9. Weinges, K., and Freudenberg, K., *Justus Liebigs Annln Chem.*, **668**, 92 (1963).
10. Freudenberg, K., and Weinges, K., *Fortschr. Chem. org. NatStoffe*, **16**, 1 (1958).
11. Hathway, D. E., and Seakins, J. W. T., *J. chem. Soc.*, 1562 (1957).
12. Hathway, D. E., and Seakins, J. W. T., *Biochem. J.*, **67**, 239 (1957).
13. Humphries, S. G., "Biosynthesis of Natural Products", (P. Bernfeld, ed.), Pergamon Press, Oxford (1963).
14. Freudenberg, K., *Experentia*, **16**, 101 (1960).
15. Freudenberg, K., and Maitland, P., *Justus Liebigs Annln Chem.*, **570**, 193 (1934).
16. Freudenberg, K., and Alonso Lama, J. M., *Justus Liebigs Annln Chem.*, **612**, 78 (1958).
17. Freudenberg, K., Stocker, J. H., and Porter, J., *Chem. Ber.*, **90**, 957 (1957).
18. Freudenberg, K., and Weinges, K., *Justus Liebigs Annln Chem.*, **590**, 140 (1954).
19. Brown, B. R., and Cummings, W., *J. chem. Soc.*, 4302 (1958).
20. Mayer, W., and Merger, F., *Justus Liebigs Annln Chem.*, **644**, 70, 79 (1961).
21. Mayer, W., and Lemke, R., *Naturwissenschaften*, **50**, 644 (1963).
22. Mayer, W., Merger, F., Frank, G., Heyns, K., and Grutzmacher, H. F., *Naturwissenschaften*, **50**, 152 (1963).
23. Freudenberg, K., Weinges, K., and Naya, U., *Naturwissenschaften*, **50**, 354 (1963).
24. Weinges, K., Naya, Y., and Toribio, F., *Chem. Ber.*, **96**, 2870 (1963); *Justus Liebigs Annln Chem.*, **681**, 161 (1965).
25. Roux, D. G., *Nature, Lond.*, **181**, 1454 (1958).
26. Balfe, M. P., Doughty, M. A., Kenyon, J., and Poplett, H. R., *J. chem. Soc.*, 605 (1942).
27. Balfe, M. P., Kenyon, J., and Thain, E. M., *J. chem. Soc.*, 386 (1957).

28. Balfe, M. P., Kenyon, J., and Thain, E. M., *J. chem. Soc.*, 790 (1952).
29. Kenyon, J., and Mason, R. F., *J. chem. Soc.*, 4964 (1952).
30. Kenyon, J., and Davies, A. G., *Q. Rev. chem. Soc.*, **9**, 203 (1955).
31. Brown, B. R., Cummings, W., and Newbould, J., *J. chem. Soc.*, 3677 (1961).
32. Freudenberg, K., "Chemistry of Flavonoid Compounds", (T. A. Geiseman, ed.), Pergamon Press, Oxford (1961).
33. Clark-Lewis, J. W., and Mortimer, P. I., *J. chem. Soc.*, 4106 (1960).
34. Hathway, D. E., *J. chem. Soc.*, 520 (1958).
35. Barton, D. H. C., Deflorin, A. M., and Edwards, O. E., *J. chem. Soc.*, 530 (1956).
36. Barton, D. H. R., *Proc. chem. Soc.*, 293 (1963).
37. Day, A. C., Nabney, J., and Scott, A. I., *Proc. chem. Soc.*, 284 (1960).
38. Matsuura, T., and Cahnmann, H. J., *J. Am. chem. Soc.*, **82**, 2050, 2055 (1960).
39. Barton, D. H. R., and Cohen, T., "Festschrifft, Arthur Stolle" *Birkhauser, Basle*, 117 (1957).
40. Davidson, T. A., and Scott, A. I., *Proc. chem. Soc.*, 390 (1960).
41. Brown, C. J., Clark, D. E., Ollis, W. D., and Veal, P. L., *Proc. chem. Soc.*, 393 (1960),
42. Kariyone, T., and Sawada, T., *J. pharm. Soc. Japan*, **78**, 1020 (1958).
43. Baker, W., Finch, A. C. M., Ollis, W. D., and Robinson, K. W., *J. chem. Soc.*, 1477 (1963).
44. Roberts, E. A. H., *J. Sci. Fd Agric.*, **9**, 385 (1958).
45. Roberts, E. A. H., and Myers, M., *J. Sci. Fd Agric.* **10**, 177 (1959).
46. Roberts, E. A. H., and Myers, M., *J. Sci. Fd Agric.*, **11**, 153 (1960).
47. Roberts, E. A. H., *J. Sci. Fd Agric.*, **9**, 212 (1958).
48. Roberts, E. A. H., Cartwright, R. A., and Oldschool, M., *J. Sci. Fd Agric.*, **8**, 72 (1957).
49. Roberts, E. A. H., and Myers, M., *J. Sci. Fd Agric.*, **10**, 167 (1959).
50. Roberts, E. A. H., and Myers, M., *J. Sci. Fd Agric.*, **11**, 158 (1960).
51. Takino, Y., Imagawa, H., Horikawa, H., and Tanaka, A., *J. Japan. agric. biol. Chem.*, **28**, 64 (1964).
52. Takino, Y., and Imagawa, H., *J. Japan. agric. biol. Chem.*, **28**, 125 (1964).
53. Takino, Y., and Imagawa, H., *J. Japan. agric. biol. Chem.*, **27**, 666 (1963).
54. Quesnel, V. C., *Tetrahedron Lett.*, **48**, 3699 (1964).

CHAPTER 4

The Hydrolysable Tannins

I. INTRODUCTION

As their name infers the hydrolysable tannins are complex polyphenolic substances which may be degraded into simpler fragments under hydrolytic conditions (a property arising from the polyester type of structure which they all possess). Subdivision of the hydrolysable tannin group is usually made[1] on the basis of the polyphenolic acid (s) liberated on hydrolysis, those yielding gallic acid (1) only are referred to as *gallotannins* and those which give ellagic acid (2) amongst the acidic products as *ellagitannins*. A greater part of the knowledge of the

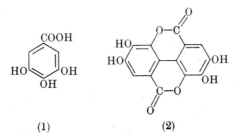

(1)   (2)

chemical structure and properties of the hydrolysable tannins has been gained as a result of a study of the extracts in this group which have or had an economic importance in commercial tannage and a selection of these is shown in Table I.

TABLE I

SOURCES OF HYDROLYSABLE TANNINS

Tannin	Source
Chinese tannin (Tannic acid)	Galls, leaves of *Rhus semialata*
Turkish tannin	Galls on wood of *Quercus infectoria* (Aleppo galls)
Sumach tannin	Leaves of *Rhus coriaria, R. typhina*
Myrobalans	Fruit of *Terminalia chebula*
Valonea	Acorn cups of *Quercus valonea*
Chestnut	Wood of *Castanea sativa*
Tara	Fruit pods of *Caesalpinia spinosa*
Divi-Divi	Fruit pods of *Caesalpinia coriaria*
Algarobilla	Fruit pods of *Caesalpinia brevifolia*
Knoppern nuts	Fruit of *Quercus pedunculata*
Pomegranate	Fruit, twigs, root, *Punica granatum*

Prior to the discovery and application of chromatographic methods of analysis work on the constitution of the hydrolysable tannins was severely limited by the lack of suitable methods for the preparation of the tannins in a homogeneous form and by the absence of methods to determine their purity. Thus from the present knowledge of the composition of the phenolic extract of sumach (*Rhus coriaria*) leaves, it is obvious that a great deal of the earlier and perhaps more recent investigations into the structure of the gallotannin it contains were carried out with mixtures containing extraneous components many of a flavonoid nature[2]. White and his collaborators[3] were the first to demonstrate that these difficulties could be met and resolved by the use of chromatographic methods of separation and analysis, and in a series of publications they showed by paper chromatographic analysis that the majority of the commercially available hydrolysable and condensed tannin extracts were in fact complex mixtures of polyphenols. Their work not only indicated that a reappraisal of many of the earlier researches on these substances was necessary but also prepared the way for further advances in this area of the chemistry of plant products.

II. PREPARATION OF PLANT EXTRACTS

The hydrolysable tannins are noted for their lability towards hydrolytic agents, being cleaved by enzymes, hot water, acids or alkali into their component parts. Like the condensed tannins they may also suffer irreversible changes when subject to heat or oxidizing agents and for these reasons alone care must be exercised in their extraction from plant materials. Many techniques have been described for the isolation of plant polyphenols but subsequent studies of the chemical properties of naturally occurring phenolic esters aided by chromatographic analysis have shown several of these methods to be unsuitable for this type of compound. Extraction with alcoholic solvents should for instance be avoided when dealing with phenolic esters whose immediate structural environment contains an aliphatic or phenolic hydroxyl group. Thus orthohydroxy depsides, such as are present in the gallotannins, are readily cleaved by alcohols (3 → 4) with the formation of

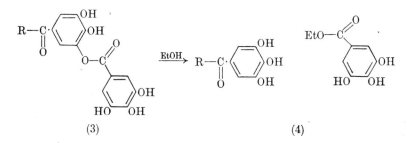

(3)   (4)

the corresponding gallate esters[4], and extraction of crushed Chinese galls with ethanol yields an extract in which ethyl gallate may be identified and from which it may be isolated. However, that this compound is an artifact produced by ethanolysis of depside linkages in the gallotannin structure may be demonstrated by extraction of the galls with aqueous acetone when the compound is absent from the extract. Under other conditions such as extraction at an elevated temperature (60–70°) water or alcohols acting as weakly basic substances may also promote, in a favourable structural environment, migration of an acyl group. Acyl migration, particularly in carbohydrate chemistry, is a well documented phenomenon and it also appears to form an important aspect of the chemistry of natural phenolic esters. Thus chlorogenic acid (5) treated in aqueous media at neutral pH at 80° for five minutes produces, by a facile acyl migration, an equilibrium mixture of the 3-, 4- and 5-O-caffeoylquinic acids (5 ↔ 6 ↔ 7)[5].

In dealing with plant galls, fruit pods and dried leaves the most satisfactory technique therefore for the isolation of the polyphenols present is to leach them from the powdered plant material by percolation or shaking with cold water or aqueous acetone and then to re-extract them from aqueous solution with cold ethyl acetate (20–30 times), finally removing the ethyl acetate at 30°. In the treatment of fresh green tissues such as stems or leaves it is usually most convenient to macerate the plant material in a blender with acetone and water, to filter off the plant debris through glass wool or a layer of cellulose and extract the resultant aqueous solution as rapidly as possible in the manner described above. Where such techniques are not possible or practicable leaves dried in air and ground to a fine powder should be used. However, many plant leaves, such as those from *Bergenia, Vaccinia* or *Pyrus* species, undergo a fairly rapid browning on exposure to air.

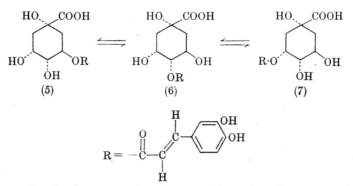

The browning in these cases has been attributed to the enzymic hydrolysis of the arbutin (8) they contain[6] and enzymic co-oxidation of the resultant hydroquinone with other phenols present to give black polymeric products. In such instances it is imperative to extract fresh undried leaves with a minimum of delay if it is desired to have an accurate picture of the range of natural phenols produced by the plant.

III. PAPER CHROMATOGRAPHIC ANALYSIS OF HYDROLYSABLE TANNINS

The methods of paper chromatographic analysis used in dealing with the condensed tannins constitute a routine method for the analysis of plant polyphenols and are readily applied in the case of the hydrolysable tannins. Unless the nature of particular components of an extract demand the use of alternative solvents it is, from several points of view, most useful to apply the standard solvent pair—(a) 6% aqueous acetate acid and (b) butan-2-ol–acetic acid–water (14:1:5)—for the analysis of the hydrolysable tannin extracts. The majority of the gallotannins and many of the ellagitannins run as discrete spots on paper chromatograms (Fig. 1) and in this respect are more amenable to satisfactory paper chromatographic analysis than the corresponding condensed tannins.

Galloyl and hexahydroxydiphenoyl esters absorb or fluoresce under ultra-violet light (both characteristics are normally enhanced by fuming in ammonia vapour), and this constitutes a useful preliminary method for their detection on paper chromatograms. Various spray reagents may also be employed. Two general but relatively unspecific sprays are the ferric chloride–potassium ferricyanide[7] and diazotized benzidine reagents[8] (see p. 24), which reveal the esters as respectively blue and orange-yellow spots, on a white background. Two rather more specific spray reagents which may be applied to detect galloyl or hexahydroxydiphenoyl esters are described below.

1. POTASSIUM IODATE SOLUTION[9]

A spray of a freshly prepared saturated solution of potassium iodate reveals the esters as pink-red spots on a white background. The colours are stable for up to 30 min when they slowly turn brown in appearance. Gallic acid rapidly changes colour from red to brown (2–3 min), due to

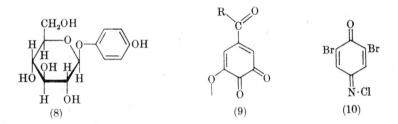

(8)   (9)   (10)

the ready formation of purpurogallin carboxylic acid. The specificity of the spray is due to production from galloyl esters under mildly oxidative conditions of hydroxyquinone derivatives such as (9), which have characteristic absorptions in the visible region of the spectrum.

2. GIBBS REAGENT[10]

A spray of a freshly prepared solution (0·5%) of 2,6-dibromoquinone-N-chlorimide (10) in methanol or acetone followed by one of saturated sodium bicarbonate solution. The test is dependent on the formation of indophenols by substances having an unsubstituted —CH— group para to a phenolic group, and has been used both for the qualitative and quantitative analysis of such compounds. Various ambiguities have been associated with the use of this reagent however, and it reacts to give colours ranging from yellow to purple with phenols many of which do not possess a free para position. Thus for example p-coumaroyl esters give a yellow, hydroquinone a purple and arbutin (8) a bright blue colour with Gibbs reagent. Galloyl and hexahydroxydiphenoyl esters are normally revealed by the spray as brown-purple areas on the chromatogram.

The reaction of these spray reagents towards a number of natural phenolic esters and related products together with their behaviour under ultra-violet light is recorded in Table II.

TABLE II

DETECTION OF SOME NATURAL PHENOLIC ESTERS ON PAPER CHROMATOGRAMS

	Spray			Ultra-violet light	
Compound	Ferric chloride-potassium ferricyanide	Potassium iodate	Gibbs reagent		Ammonia vapour
Sumach gallotannin	+	Pink-red	Brown	Absorption	Absorption
Acer tannin	+	Pink-red	Brown	Violet*	Violet*
Corilagin	+	Pink	Brown	White-blue*	White-blue*
β-penta-O-galloyl-D-glucose	+	Pink	Brown	Absorption	Violet*
3-O-p-coumaroyl quinic acid	−	—	Yellow	Weak Violet*	Blue*
3-O-feruloyl quinic acid	−	—	Yellow	Blue*	Blue-green*
3-O-caffeoyl quinic acid	+	Yellow-brown	Brown	Blue*	Green*
Gallic acid	+	Pink	Brown	Violet*	Violet*
Ellagic acid	+	—	Brown	White-blue*	White-blue*

The structural features which determine the R_f values of particular galloyl or hexahydroxydiphenoyl esters in the solvent systems recommended are essentially analogous to those enumerated previously in a discussion of the analysis of flavonoid compounds. The most prominent feature is that an increase in the number of galloyl groups in the molecule leads to a considerable decrease in the R_f value in the first phase of development and concommittantly that an increase in the number of aliphatic hydroxyl groups increases the R_f value. Differences in the content of alcoholic or phenolic hydroxyl groups do not have such a marked effect on the R_f values of compounds in the organic partitioning solvent; thus for β-D-glucogallin (11) and β-penta-O-galloyl-D-glucose (12) the ΔR_f in butan-2-ol–acetic acid–water (14:1:5), is 0·28 compared to a value of 0·67 in the 6% aqueous acetic acid phase. An examination of Table III in which the chromatographic properties of a number of phenolic esters are listed reveals the generality of these trends.

Detailed examination of the paper chromatogram of a hydrolysable tannin extract normally reveals compounds whose presence undoubtedly lends particular properties to the extract but which individually

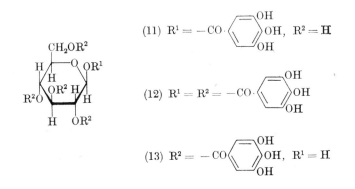

(11) $R^1 = -CO \cdot$ ⟨OH, OH, OH⟩, $R^2 = H$

(12) $R^1 = R^2 = -CO \cdot$ ⟨OH, OH, OH⟩

(13) $R^2 = -CO$ ⟨OH, OH, OH⟩, $R^1 = H$

cannot be classified as tannins. Typical of this type of compound are the flavonoids, coumarins, stilbenes, esters and glucosides of the hydroxycinnamic acids and other glucosides such as arbutin and its derivatives. Methods for the recognition of the flavonoids and stilbenes have been outlined in chapter two and relevant characteristics of some of the other compounds are listed in Tables II and III. The hydroxycinnamoyl esters are most readily distinguished by their blue-violet fluorescence in ultra-violet light and arbutin and its derivatives by their distinctive blue colouration with Gibbs reagent. Since the effect of these compounds on the process of tannage is as yet undetermined their identification remains an important facet of vegetable tannin chemistry. In addition it is hoped that a fuller understanding of the metabolic pathways involved in plant polyphenol formation will be assisted by the characterization of many of these simpler but intrinsically important molecules.

4*

TABLE III

R_f VALUES OF SOME PHENOLS, PHENOLIC ESTERS AND
GLYCOSIDES

Compound	R_f values 6% acetic acid	B.A.W.*
Gallic acid	0·52	0·60
m-Digallic acid	0·25	0·73
Methyl gallate	0·62	0·72
Methyl-m-digallate	0·37	0·86
Ellagic acid	0·01	0·35
2-O-Galloylglycerol	0·56	0·46
1-O-Galloylglycerol	0·64	0·51
1,2-Di-O-galloylglycerol	0·13	0·41
1,3-Di-O-galloylglycerol	0·23	0·47
Tri-O-galloylglycerol	0·06	0·54
Mono-O-galloyl-cyclohexane-cis-1,2-diol	0·63	0·79
Mono-O-galloyl-cyclohexane-trans-1,2-diol	0·59	0·85
β-1-O-Galloyl-D-glucose (β-glucogallin)	0·75	0·30
3,6-Di-O-galloyl-D-glucose	0·45	0·42
3,4,6-Tri-O-galloyl-D-glucose	0·33	0·38
2,3,6-Tri-O-galloyl-D-glucose	0·25	0·56
2,3,4,6-Tetra-O-galloyl-D-glucose	0·21	0·60
β-Penta-O-galloyl-D-glucose	0·08	0·58
Corilagin	0·35	0·20
β-1-O-p-Coumaroyl-D-glucose	0·75	0·75
β-1-O-Caffeoyl-D-glucose	0·70	0·50
β-1-O-Feruloyl-D-glucose	0·72	0·62
Hydroquinone-β-D-glucoside (Arbutin)	0·95	0·50
1-O-Galloylquinic acid	0·69	0·34
1-O-Galloylquinide	0·54	0·62
Ethyl-1-O-galloylquinate	0·76	0·58
3-O-Galloylquinic acid	0·62	0·46
3,4,5-Tri-O-galloylquinic acid	0·35	0·52
1,3,4,5-Tetra-O-galloylquinic acid	0·22	0·42
3-O-p-Coumaroylquinic acid	0·69	0·78
3-O-Caffeoyl quinic acid (chlorogenic acid)	0·60	0·70
3-O-Feruloylquinic acid	0·66	0·72
5-O-Caffeoylquinic acid (neochlorogenic acid)	0·69	0·57

* B.A.W.—Butan-2-ol–acetic acid–water (14 : 1 : 5).

IV. QUANTITATIVE ESTIMATION OF GALLOYL ESTERS IN PLANT EXTRACTS

The quantitative estimation of total phenols in a plant extract may be accomplished in a number of ways, many of which have been described in detail in the literature. Most of these methods are dependent either on the susceptibility of phenols to oxidation (Folin–Dennis phosphomolybdate reagent[12] and the Loewenthal potassium permanganate titration[13]), or to coupling with electrophilic species (diazotized p-nitroaniline,[14] or other aromatic amines, and Gibbs reagent[11] (10)) or to the formation of distinctively coloured complexes with transition metal salts[15] (ferrous ammonium tartrate). Phenols also show a characteristic absorption in the ultra-violet due to the aromatic nucleus and this property has been used in some types of estimation[16]. When it is necessary to estimate the amount of a particular type of phenol in an extract rather more specific reactions are, however, required. The leucoanthocyanins are for instance readily estimated[17] by means of their facile conversion with acid to highly coloured anthocyanidins which may then be determined colorimetrically.

Routine estimation of galloyl esters in plant extracts is a necessary adjunct to biosynthetic and metabolic studies of these compounds. Several procedures have been described and two which are quite specific for this class of compound are mentioned here. *Tannase* is an enzyme derived from the growth of *Aspergillus niger* on tannin media[18] and acts specifically to hydrolyse galloyl esters. The latter may therefore be quantitatively assessed in an extract by treatment of an aliquot with *tannase* to liberate gallic acid which is then estimated spectrophotometrically or titrimetrically[19]. The method is easily adapted to a microscale when the gallic acid is estimated with an autotitrator. An alternative method[20], suitable for plant tissues or extracts containing a relatively high concentration of galloyl esters, depends on the mild oxidation of galloyl groups using potassium iodate to hydroxyquinones of the type (9) whose red colour enables estimation by the normal methods of colorimetry. The hydroxyquinone derivatives (9) are fully developed and relatively stable over a period of 45 min at 0°, but at room temperature their stability is considerably decreased. A brief description of the experimental method as employed in the determination of galloyl esters in the leaves of *Acer saccharinum* is given below.

"Leaves (approximately 3 g) were crushed in a mortar with powdered glass or carborundum chips and water (5 ml) to give a fine paste. Water (50 ml) was added and the resulting suspension filtered after 5 min through a pad of iron free cellulose (10 g) in a Hirsch funnel. The pad was washed with four further quantities of water (20 ml), the aqueous extract acidified with N-sulphuric acid (1–2 ml) and extracted with ethyl acetate (8 × 100 ml). After removal of the ethyl acetate at 30° the residual gum was dissolved in acetone (2 × 10 ml) and water (15 ml) and made up to 100 ml bulk. Aliquots (0·5, 1·0, 1·5, 2·0 and 2·5 ml) were taken and added to test tubes containing respectively (3·0, 2·5, 2·0, 1·5 and 1·0 ml of H_2O) and the tubes immersed in an ice bath for 30 min. Potassium iodate solution (1·5 ml, saturated) was added to each tube and after 40 min at 0° the optical density at 550 mμ determined. The number of equivalents of gallic acid present as a galloyl ester in the test solution was then determined graphically by reference to similar measurements on a solution of β-penta-O-galloyl-D-glucose (100 mg) in acetone (20 ml) and water (80 ml)."

V. ISOLATION OF THE HYDROLYSABLE TANNINS

In several cases where the desired galloyl or hexahydroxydiphenoyl derivative is readily crystallized and is present in a plant extract in high concentration, accompanied by few extraneous compounds, direct crystallization may be a possible method of isolation. Acer[21] and Hamameli[22] tannins, gallic, ellagic and chebulinic acids[23] are several of the compounds which have been isolated in this way. Other crystalline components of tannin extracts are often readily accessible after an initial fractionation (e.g. lead salt precipitation, counter-current distribution) and using a combination of these techniques Schmidt and his collaborators have obtained corilagin[24], brevifolin carboxylic acid[25], chebulagic acid[26] and other important compounds of the ellagitannin class in a crystalline state. However, for the more complex derivatives such as the gallotannins which are amorphous even when prepared in a homogeneous state, as judged by paper chromatographic analysis, other methods of isolation are necessary. Chinese and Turkish gallotannins are conveniently prepared[2] by ethyl acetate extraction of a buffered (pH 6·8) solution prepared from the crushed galls of *Rhus semialata* and *Quercus infectoria* respectively. Tara gallotannin[27] (fruit pods of *Caesalpinia spinosa*) is best purified by counter-current distribution between methyl ethyl ketone and water and by precipitation from ethyl acetate solution with benzene. Further fractionation of the tannins is then possible by chromatography on cellulose using 6% acetic acid as eluant. Sumach gallotannin (leaves *Rhus typhina, Rhus coriaria*) is conveniently isolated and freed from material of a flavonoid nature by chromatography on polyamide powder[2]. Elution at 0° with methanol removes the greater part of the flavonoid material and the tannin may then be obtained using acetone or methyl ethyl ketone as eluant. Once isolated these tannins are most conveniently handled for subsequent chemical degradation and analysis in a freeze-dried state (from water, acetic acid or 1,1-dimethylethanol).

As is the case in the isolation of flavonoid compounds from condensed tannin extracts the relative disposition of compounds on paper chromatograms of hydrolysable tannin extracts may be employed as a reliable guide to the methods which may be used for the isolation of

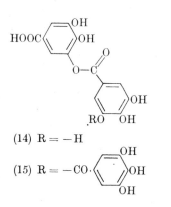

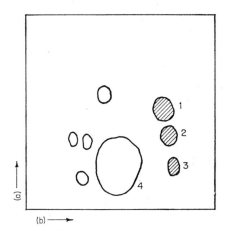

(14) R = −H

(15) R = −CO·

FIG. 1. Paper chromatogram of extract of Chinese galls. (a) 6% aqueous acetic acid (b) butan-2-ol–acetic acid–water (14:1:5). 1, Gallic acid; 2, *m*-digallic acid; 3, *m*-trigallic acid; 4, Chinese gallotannin.

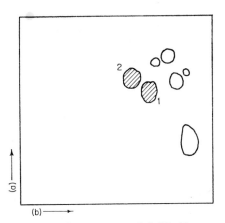

FIG. 2. Paper chromatogram of an extract of Californian prunes (a) 6% aqueous acetic acid; (b) butan-1-ol–acetic acid–water (14:1:5). 1, chlorogenic acid; 2, neo-chlorogenic acid.

many of the minor constituents. Compounds which are separated in the aqueous phase of the development of the paper chromatogram may in general be isolated by the application of large scale cellulose column or thin layer cellulose chromatography and an excellent example of this is in the separation of gallic (1), *m*-digallic (14) and *m*-trigallic (15) acids from the acidic fraction of tannic acid (Chinese galls)[28]. The paper chromatogram of this extract is depicted in Fig. 1 and shows clearly the separation of these three components in the first phase of development. Analogously compounds which have comparable R_f values in the aqueous phase of paper chromatographic analysis but have different values in the organic phase are usually separable by counter-current distribution. An example of the practise of this method is the isolation of chlorogenic (5) and neochlorogenic (7) acids from admixture in a plant extract (Fig. 2). The small difference in R_f value between these compounds in the organic partitioning phase (Fig. 2) is readily exploited in a counter-current separation of the two acids between ethyl acetate and water[5] (Fig. 3).

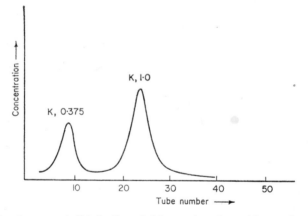

FIG. 3. Counter-current distribution of chlorogenic and neochlorogenic acids (K, 1·0 and 0·375 respectively) between ethyl acetate and water (equal phase volume, 50 transfers).

VI. ANALYSIS OF HYDROLYSABLE TANNINS

The hydrolysable tannins, particularly those which are obtained in an amorphous state, retain water tenaciously and for correct elementary analyses rigorous drying (usually in a high vacuum at 100° for 1–2 days) is necessary. Qualitative and quantitative analysis of those tannins which contain gallic acid in respect of the component parts gallic acid/carbohydrate or other fragment has been assisted enormously by use of the tannin decomposing enzyme *tannase* produced by a few moulds (*Aspergillus niger* and *Penicillium glaucum*) when grown on

Czapek's medium containing 2% tannic acid instead of sucrose[18]. Extracts of these moulds contain enzymes with galloyl esterase and carbohydrase activity both of which may be assumed as being involved in the degradation of tannins. Freudenberg, who produced *tannase* by growth of *Aspergillus niger* on an extract of Myrobalan nuts[29], first recognized the potentialities of this enzyme in this field and used tannase in both the qualitative and quantitative analysis of Hamameli tannin[22], but Dyckerhoff and Armbrüster[30] were the first workers to make a careful study of the specificity of the enzyme. Hydrolysis of methyl gallate was shown to follow pseudo-first-order kinetics and a pH optimum of 5·0 was observed for the enzyme. Using a variety of substrates Dyckerhoff and Armbrüster showed that for hydrolysis by the galloyl esterase the aromatic acid portion of the ester should contain

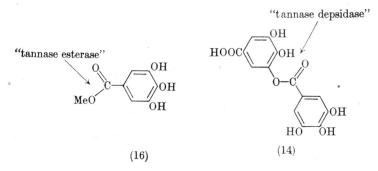

(16)   (14)

two phenolic hydroxyl groups neither of which should be ortho to the carboxyl group. Thus methyl gallate, protocatechuate and 3,5-di-hydroxybenzoate but not methyl salicylate, 2,4-dihydroxybenzoate, or 2,5-dihydroxybenzoate acted as typical substrates for the esterase. The later work of Haworth and his group[19] supported many of these findings although they were unable to confirm the suggestion of Dyckerhoff and Armbrüster regarding the presence of a mould esterase, capable of hydrolysing esters of acetic acid, in *tannase*. Toth and Barsony[31] and later Toth[32] produced evidence that chromatography over alumina resolved *tannase* into two separate esterases, one (esterase) specific for the normal aliphatic ester linkage such as in methyl gallate (16) and the other (depsidase) which hydrolysed depside links such as in *m*-digallic acid (14). Subsequent work has been unable to support these conclusions, kinetic evidence in fact suggests that *tannase* contains a single esterase with two relative specificities one for the methyl gallate type of linkage, the other for depside linkages.

Tannase, which has been obtained in a crystalline form from *Aspergillus niger*[33] and from the fruit pods of Divi-divi (*Caesalpinia coriaria*)[34], is a fairly robust enzyme and loses little activity on heating to 70° for an hour, although substantial losses occur on exposure to ultra-violet light. Haworth and his collaborators have used tannase extensively in the quantitative analysis of the gallotannins[2], the gallic acid content being determined after tannase hydrolysis spectrophotometrically or using an autotitrator, the glucose, after removal of gallic acid from the hydrolysate, by the colorimetric method of Park and Johnson[35] using anthrone.

The carbohydrase and esterase activities of tannase may be fractionated by chromatography on a basic ion exchange resin and Haworth, Haslam, Jones and Rodgers[19] obtained by this means a galloyl esterase fraction which contained no significant carbohydrase activities. Using this enzyme fraction they were able to show that the gallotannins from sumach leaves, Chinese and Aleppo galls were based on a glucose and not an oligosaccharide core, as suggested by other workers[36, 37].

An important measurement in structural studies of the gallotannins is the molecular weight, but such information of a quantitative nature which is available regarding the state of tannin molecules in solution suggests that the solutions are colloidal and polydisperse containing a range of particle sizes[36]. Several results of an analytical nature can indeed only be satisfactorily interpreted on the basis of strong association between polygalloyl esters in aqueous solution[38]. Thus β-penta-O-galloyl-D-glucose (12) and 2,3,4,6-tetra-O-galloyl-D-glucose (13) gave Maxwellian distribution curves (K, 1·04 and 1·34 respectively) when subject to counter-current distribution in the two phase quaternary system propan-1-ol–butan-1-ol–cyclohexane–water (33:11:7:49), but an equimolar mixture of the two esters when similarly analysed gave an exact Maxwellian distribution (K, 1·19). Similar results also obtained for the counter-current analysis of the mixture of isomeric tetra-O-galloyl-D-glucose esters resulting from methanolysis of Turkish gallotannin. Measurements of the molecular weights of many of the hydrolysable tannins, particularly in aqueous solution, are therefore hindered by the tendency of these compounds to associate to form higher molecular weight aggregates. Thus at 1% concentration in aqueous solution the molecular weight of Chinese gallotannin measured in the ultra centrifuge (sedimentation equilibrium) and in the thermoelectric osmometer was 2500, indicative on the basis of other analytical date, of a dimeric form. In more concentrated solution the molecular weight increased to the region of 4–5000. Similar results were obtained for other gallotannins under identical conditions. However, in acetone solution the tannins exist predominantly in the monomeric state, since synthetic β-penta-O-galloyl-D-glucose gave a value using the osmometer of 944 (theoretical 940) and under similar circumstances Chinese gallotannin gave a value of 1253 indicative of either a hepta or an octagalloyl glucose structure. Cryoscopic measurements of molecular weight determination have been employed by other workers thus in acetic acid Frank and Freudenberg[39] found a value of 895–956 for chebulinic acid and for Chinese gallotannin[40] in dimethyl succinate a value of 1690–1793. On the whole however data concerning particle form and size in tannin solutions is only fragmentary and a systematic study would be of undoubted value, not the least in relation to the mechanism of tannage.

VII. GALLOTANNINS AND OTHER NATURALLY OCCURRING GALLOYL ESTERS

Naturally occurring galloyl esters vary in complexity from the simple mono and di-esters such as β-D-glucogallin (11), theogallin (17), (−)-epicatechin and (−)-epigallocatechin gallates (18 and 19), Hamameli

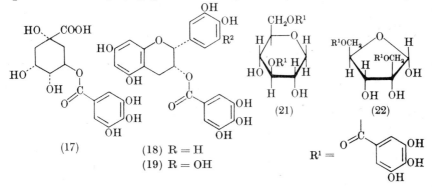

(17)

(18) R = H
(19) R = OH

(21)

(22)

and Acer tannins (22 and 21) to the gallotannins—Chinese, Sumach, Tara and Turkish—themselves, and are shown in Table IV. Galloyl esters such as corilagin, chebulagic and chebulinic acids are excluded from Table IV since they are more readily considered in relation to the ellagitannins. Similarly although a variety of other naturally occurring galloyl esters (e.g. from *Geranium*, *Arctostaphylos*, *Acer* sp.[20] and *Paeony*[41]) appear to be related in properties to one or other of the galloyl esters recorded in Table IV as their structures remain unsubstantiated they are not included.

Apart from the gallotannins only β-penta-*O*-galloyl-D-glucose (12) of the other galloyl esters shown in Table IV complies with the requirements laid down by White[36] for tanning action—namely a polyphenolic molecule with a molecular weight in the range 500–3000. Indeed Perkin and Uyeda[21] specifically pointed out that Acer tannin

TABLE IV

NATURALLY OCCURRING GALLOYL ESTERS

Compound	Source
β-D-Glucogallin (11)	Roots *Rheum officinale*,[42] Myrobalans
Theogallin (17)	Leaves *Camellia* sp.[43]
Epicatechin gallate (18)	Leaves *Camellia* sp.[43, 44]
Epigallocatechin gallate (19)	Leaves *Camellia* sp. [43, 44]
3,6-Di-*O*-galloyl-D-glucose	Myrobalans[45]
β-1,3,6-Tri-*O*-galloyl-D-glucose	Myrobalans[45]
Acer tannin (21)	Leaves of *Acer* sp.[21]
(3,6-di-*O*-galloyl-1,5-anhydro-D-glucitol)	
Hamameli tannin (22)	Bark of *Castanea* and *Hamamelia*
(2′,5-di-*O*-galloyl-D-hamamelose)	sp.[22]
β-Penta-*O*-galloyl-D-glucose (12)	Myrobalans[45]
Chinese, Dhava, Sumach gallotannins	Galls *Rhus semialata*[2], leaves
(hepta-nona-*O*-galloyl-D-glucose)	*Anogeissus latifolia*[46], leaves
	Rhus typhina[2], *R. coriaria*[2]
Turkish gallotannin	Aleppo galls (*Quercus infectoria*)[2]
(hexa-hepta-*O*-galloyl-D-glucose)	
Tara gallotannin	Fruit pods (*Caesalpinia spinosa*)[27]
(tetra-penta-*O*-galloyl-quinic acid)	

has little if any tanning action and hence its classification (and probably that also of Hamameli tannin) as a tannin is in this sense incorrect. The qualitative experiments of Russell and Tebbens[47] present useful guides to the specific requirements for tanning properties in a series of galloyl esters. Simple esters of gallic acid (methyl, ethyl, *n*-hexyl) they found had no tanning action but esters with various polyalcohols had tanning properties that were initially poor but which improved as the alcohol homologous series was ascended. Thus ethylene glycol digallate had poor tanning characteristics but mannitol and sorbitol hexagallates had a fair tanning ability. Excluding β-penta-*O*-galloyl-D-glucose which Russell and Tebbens[47] similarly reported as having a good capacity for tannage of calf skin, the naturally occurring galloyl esters which act as tannins are Chinese, Sumach, Turkish and Tara gallotannins and it is perhaps significant in relation to the mechanism of tannage that all these substances contain within their structures some gallic acid bound in a depside form.

1. CHINESE, SUMACH AND TURKISH GALLOTANNINS

The gallotannins obtained from plant galls—on the leaves of *Rhus semialata* (Chinese gallotannin or tannic acid), and the wood of *Quercus infectoria* (Turkish gallotannin)—occupy an important position amongst the vegetable tannins. Not only were they commercially valuable in earlier times but from a historical point of view these tannins were those whose chemical structures were first subject to detailed investigation in the period 1910–1930 by a series of distinguished chemists including Emil Fischer[48], Freudenberg[40] and Karrer[49, 50]. Although the introduction of paper chromatographic techniques has indicated that these and other tannin extracts are mixtures of a somewhat more complex nature than these workers appreciated, from this early work there emerged a picture of the type of structure involved in the gallotannin molecules[40, 48] and in the case of Chinese gallotannin one which later work has shown to be remarkably accurate[2].

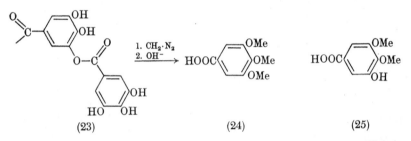

(23)   (24)   (25)

The work of Fischer, Freudenberg and their collaborators on Chinese and Turkish gallotannins showed them to be polygalloylglucose derivatives in which the ratio of glucose to gallic acid was 1 to 9 or 10 and 1 to 5 or 6 respectively[48]. In both gallotannins a certain percentage of the galloyl groups was shown to be bound in the form of depsides (23) by the isolation of both 3,4,5-tri-*O*-methyl gallic (24) and 3,4-di-*O*-methyl-gallic (25) acid after hydrolysis of the diazomethane methylated tannins. In the case of Chinese gallotannin the ratio of these acids was approximately 1 : 1 and Fischer was therefore led to suggest that in this substance all the hydroxyl groups on the sugar molecule were esterified with gallic acid and that the tannin had a composition approximating to that of β-penta-*m*-digalloyl-D-glucose (26). In support of this suggestion a synthetic specimen of the latter (prepared by condensation of β-D-glucose and penta-*O*-acetyl-*m*-digalloyl chloride (29) followed by removal of the acetyl groups with alkali[51]), showed many similarities in chemical and physical properties to the natural tannin. Fischer, however, pointed out[48] that with the available evidence there was some justification for assuming that the gallotannin was a mixture of closely related substances since even if the plant gall produced a compound of the composition of β-penta-*m*-digalloyl-D-glucose there would be sufficient opportunity during its isolation for the partial hydrolysis of galloyl groups by enzymic or other means. Fischer also envisaged[48] the possibility that the tannin molecule might contain accumulations of galloyl groups in the form not only of digalloyl but also of tri and tetra-galloyl chains although he did not rate this possibility very high. Many of these ideas were later summarized by Freudenberg[40] who pointed out that with the evidence then available approximately 200 structures for Chinese gallotannin varying from β-penta-m-digalloyl-D-glucose (26) to (27) were compatible with the experimental evidence. Support for the idea that Chinese gallotannin was a heterogeneous mixture of substances

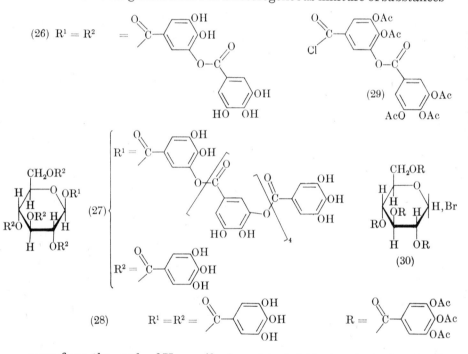

came from the work of Karrer[49] who subjected the tannin to fractional precipitation with alumina and obtained fractions of differing optical rotation which with acetic and hydrobromic acids gave varying amounts of tetrakio-tri-O-acetylgalloyl-D-glucosyl bromide (30).

The views of Fischer[48] and Freudenberg[40] and later Karrer[50] regarding the structure of Turkish gallotannin were less clearly defined. Feist[52] in 1912 had pointed out that significant differences existed between the extracts of Chinese and Turkish (Aleppo) galls; the latter extract in particular was more evidently a mixture and was character-ized by the presence of large quantities of ellagic acid. Fischer found that Turkish tannin had many similarities to β-penta-O-galloyl-D-glucose (28) but the isolation of small quantities of 3,4-di-O-methylgallic acid (25) from the hydrolysis of the methylated tannin indicated that some of the galloyl groups in the tannin were bound in depside form (23). Fischer and Freudenberg[53] also favoured the view that the large quantities of ellagic acid present in the extract arose from a water soluble glucoside distinct from the tannin but Karrer[50] suggested that it replaced some of the gallic acid in the tannin structure. Karrer[50] also produced some evidence to show that not all the hydroxyl groups on the glucose core were esterified by gallic acid and in 1933 Freudenberg[40] summarized the position in relation to the structure of this tannin by concluding that on average one of the five glucose hydroxyl groups was unesterified, one was esterified by m-digallic acid (14) and the remainder with gallic acid.

The structure of Sumach gallotannin was also far from clear during this initial period of interest in the tannins. Several workers[54] proposed a relationship to Chinese gallotannin but others including Karrer[50] implied structural similarities to Turkish gallotannin. More recent work on the gallotannins, in particular by the Sheffield school[2, 55], has confirmed the general features of the structures of Chinese and Turkish gallotannins suggested by Fischer and Freudenberg and clarified many of the finer details. The identity of Chinese and Sumach gallotannins has been substantiated[2] and claims which were made for the existence of tri- and tetrasaccharide cores in these tannins have been shown to be erroneous. The tannins, isolated as outlined previously, were obtained as freeze dried amorphous powders. Comparison of the relevant properties (elementary analysis, optical rotation, infra-red spectra, gallic acid and glucose contents) demonstrated the identity of Chinese and Sumach gallotannins as octa or nona galloylglucose derivatives[2] but the divergences apparent with Turkish gallotannin (a hexa or hepta galloyl glucose) suggested structural differences in this case[55].

In principal the steps in the structural investigation of the gallotannins by Haworth and his colleagues involved an examination of three features; (i) the nature of the carbohydrate fragment (ii) the extent of esterification of the latter by gallic acid to give the *core* of the tannin and (iii) the positions of attachment to this *core* of the depsidically linked galloyl groups. The constitution of the carbohydrate fragment was confirmed for all these tannins as being D-glucose by the action of the carbohydrase free galloyl esterase obtained from tannase by ion exchange chromatography[19]. Similar conclusions also obtained from *trans*-esterification of the diazomethane methylated tannins (sodium methoxide in methanol) when methyl 3,4,5-tri-O-methyl, and

3,4-di-*O*-methylgallate and D-glucose were the sole solable products[2]. In the face of this evidence claims for the presence of an oligosaccharide core in Chinese[36] and Sumach gallotannins[37] were withdrawn[2, 56].

Evidence of the extent of esterification by gallic acid of the glucose nucleus was obtained by hydrolysis or transesterification of the permethylated tannins (diazomethane, followed by methyl iodide and silver oxide). Chinese and Sumach gallotannins gave in this reaction sequence D-glucose as the only identifiable carbohydrate[2] but Turkish gallotannin gave a mixture of two monomethyl-D-glucose derivatives which paper electrophoresis in borate and germanate[57] buffers showed to be a mixture of 2- and 4-*O*-methyl-D-glucose[55]. These results therefore indicated that the structure of Chinese and Sumach gallotannins is based predominantly on a fully esterified penta-*O*-galloyl-D-glucose

$$(31)\ R^1 = H;\ R^2 = R^3 = R^4 = R^6 = \begin{array}{c} O \\ \| \\ C- \end{array} \hspace{-0.5em}\bigcirc\hspace{-1em}\begin{array}{l} OH \\ OH \\ OH \end{array}$$

$$(32)\ R^1 = R^2 = R^3 = R^4 = R^6 = \begin{array}{c} O \\ \| \\ C- \end{array} \hspace{-0.5em}\bigcirc\hspace{-1em}\begin{array}{l} OMe \\ OMe \\ OMe \end{array}$$

$$(33)\ R^1 = R^2 = R^3 = R^4 = R^6 = \begin{array}{c} O \\ \| \\ C- \end{array} \hspace{-0.5em}\bigcirc\hspace{-1em}\begin{array}{l} OCH_2 \cdot C_6H_5 \\ OCH_2 \cdot C_6H_5 \\ OCH_2 \cdot C_6H_5 \end{array}$$

$$(34)\ R^1 = H;\ R^2 = R^3 = R^4 = R^6 = \begin{array}{c} O \\ \| \\ C- \end{array} \hspace{-0.5em}\bigcirc\hspace{-1em}\begin{array}{l} OCH_2 \cdot C_6H_5 \\ OCH_2 \cdot C_6H_5 \\ OCH_2 \cdot C_6H_5 \end{array}$$

core and that of Turkish gallotannin on a mixture of galloylated glucose cores having unesterified hydroxyl groups at positions 2 and 4 severally. Confirmation of these deductions came from the application of a novel reaction[4] in which galloyl groups bound depsidically in the gallotannins were cleaved from the core of the tannin by the action of methanol at neutral pH. After 7 days' methanolysis of Chinese and Sumach gallotannins followed by chromatography of the products on cellulose—methyl gallate, β-penta-*O*-galloyl-D-glucose (28) and 2,3,4,6-tetra-*O*-galloyl-D-glucose (31) were isolated. Confirmation of the structure, β-configuration of the glucosidic galloyl group and pyranose ring form of the glucose in β-penta-*O*-galloyl-D-glucose was obtained from several sources. Methylation with diazomethane gave the crystalline β-pentakis-tri-O-methylgalloyl-D-glucose (32) identical[2] with the product of condensation of β-D-glucose and tri-O-methylgalloyl chloride[53]. The β-configuration at C_1 in this synthetic product had been earlier demonstrated by Fischer and Freudenberg[53] by its difference from the condensation product of tri-O-methylgalloyl chloride and α-D-glucose and hence the penta-O-galloyl-D-glucose (28) isolated from the methanolysis of Chinese gallotannin must also have the β-configuration at the glucosidic carbon atom. Condensation of tri-O-benzylgalloyl chloride with β-D-glucose followed by hydrogenation of the product (33) yielded a compound identical[2] in all respects with (28) and which on methylation gave the crystalline (32), further substantiating these structural assignments.

The minor galloyl glucose derivative obtained from the methanolysis of the Chinese and Sumach gallotannins gave a positive reaction with aniline hydrogen phthalate and its structure as 2,3,4,6-tetra-O-galloyl-D-glucose (31) was confirmed by synthesis. Treatment of β-pentakis-tri-O-benzylgalloyl-D-glucose with acid washed alumina gave a crystalline substance formulated on the basis of its reducing properties as (34) and hydrogenation gave a substance identical to (31). The β-D-glucopyranose configuration of (28) was further substantiated[38] by the reaction sequence shown in Fig. 4. Methylation of (34) gave the non reducing β-methyl-2,3,4,6-tetrakis-tri-O-benzylgalloyl-D-glucoside (35) identical with the product of condensation of tri-O-benzylgalloyl chloride and β-methyl-D-glucoside thus confirming the pyranose ring form of degradation products (28) and (31). Haworth and his collaborators suggested[2] that the small amounts of the tetra-O-galloyl-D-glucose derivative (31) formed during the anaerobic methanolysis of Chinese and Sumach gallotannins probably arose as a result of a partial breakdown of the tannin before or during its isolation. Some support for this idea was obtained by extraction of Sumach gallotannin from freshly plucked leaves under carefully controlled conditions; subsequent methanolysis gave β-penta-O-galloyl-D-glucose (28) with negligible amounts of the tetra-O-galloyl derivative. Haworth and his colleagues were thus able to conclude that Chinese and Sumach gallotannins were mixtures of closely related substances many of which were probably isomeric having β-penta-O-galloyl-D-glucose (28) as their basic core, although this pattern might be modified by changes in galls or older leaves or during isolation.

Application of the methanolysis reaction to Turkish gallotannin gave[55] methyl gallate, a non reducing tetra-O-galloyl-D-glucose and in much smaller amounts two reducing tri-O-galloyl-D-glucose derivatives all of which were isolated by a combination of chromatography on

cellulose and counter-current distribution. Methylation studies analogous to those performed on the tannin itself enabled the tetra-O-galloyl-
D-glucose to be formulated as a mixture of the 1,3,4,6- and 1,2,3,6-
esters (36) and (37) (which proved to be inseparable) and the two minor
components as the corresponding 3,4,6- and 2,3,6-tri-O-galloyl-D-
glucoses (38) and (39). These latter two minor components probably

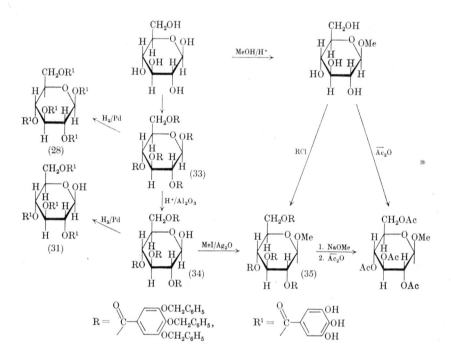

Fig. 4. Structure of the methanolysis products of Chinese and Sumach gallotannins.

arise in an analogous way to the minor products in the methanolysis
of Chinese and Sumach gallotannins and by similar arguments it was
concluded that Turkish gallotannin was a mixture of closely related
substances having a basic galloyl glucose core of (36) and (37) in an
unknown ratio.

The disposition of the remaining depsidically linked galloyl groups
on the basic cores of Chinese, Sumach and Turkish gallotannins remains
a point very much open to discussion. The distribution may be random
and this would support the view which attempts at fractionation of the
gallotannins tend to emphasize, namely that these substances are

mixtures of an extremely complex nature. Alternatively the view that these additional galloyl groups are located in the form of a polygalloyl chain at one position on the tannin core is supported by a number of lines of evidence. Thus acid hydrolysis of Chinese gallotannin yields *m*-trigallic acid (15) in a significant amount [58] and interruption of the methanolysis of all the gallotannins at an intermediate stage yields a mixture of products from which methyl-*m*-digallate can be isolated. Both of these observations are only consistent with the view that some and perhaps all of the gallotannin molecules contain a chain of three

(36) $R^2 = H$; $R^1 = R^3 = R^4 = R^6 =$

(37) $R^4 = H$; $R^1 = R^2 = R^3 = R^6 =$

(38) $R^1 = R^2 = H$; $R^3 = R^4 = R^6 =$

(39) $R^1 = R^4 = H$; $R^2 = R^3 = R^6 =$

galloyl groups.† Further more definite work is obviously required to finalize this important facet of their structures.

2. TARA GALLOTANNIN

The tara shrub (*Caesalpinia spinosa*) is a leguminous plant native to the northern part of South America and its fruit pods, which vary in colour from pale yellow to red, when crushed give the tara powder of commerce. Tara extracts are notable for their high acidity and the reasons for this particular property are due to the unique chemical structure of the gallotannin which the extract contains. Thus on mild acid hydrolysis the tannin gave gallic acid and instead of the usual carbohydrate fragment (e.g. D-glucose) the alicyclic acid quinic acid (40)[27, 59] and the acidity of the tannin is directly related therefore to the presence in its structure of the free carboxyl group of quinic acid. The tannin was readily hydrolysed by tannase and thus had the structure of a polygalloyl ester of quinic acid, analysis suggesting a tetra- to

† Recent work has provided definitive evidence for this view.

penta-*O*-galloylquinic acid[27, 59]. On methanolysis it gave 3,4,5-tri-*O*-galloylquinic acid (41) the structure of which was proved by synthesis[60] and degradation (Fig. 5). Quinic acid (40) was converted by the sequence of reactions shown to diphenylmethyl-1-*O*-benzyl-quinate (42) which

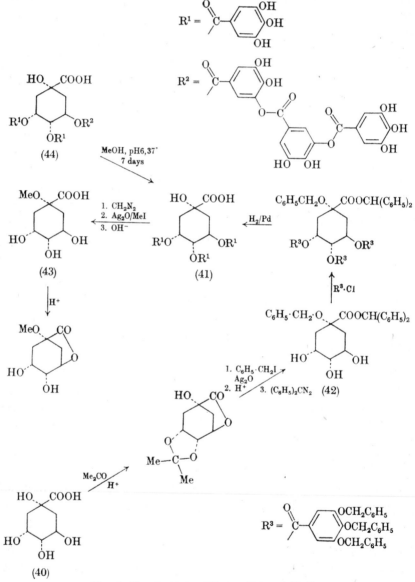

FIG. 5. The structure of Tara gallotannin[59, 60].

after condensation with tri-O-benzylgalloyl-chloride and hydrogenation of the product gave[60] the ester (41) identical in all respects with the compound from methanolysis. This structural assignment was confirmed by hydrolysis of the permethylated tannin (diazomethane followed by silver oxide, methyl iodide) when 1-O-methylquinic acid (43) was isolated and characterized. As with the other gallotannins methyl m-digallate was identified amongst the products at an intermediate stage in the methanolysis reaction thus favouring the view that some at least of the molecules of Tara gallotannin contain a trigalloyl chain. However, until additional experimental evidence is available which defines more accurately the position of the depsidically linked galloyl groups on the core the structure (44), although in good agreement with the degradative data quoted, may represent only one of several isomeric structures which contribute to the structure of Tara tannin[27].

3. ACER AND HAMAMELI TANNINS

The use of the term tannin to describe both of these digalloyl carbohydrate esters isolated from maple species and the witch hazel respectively is certainly inconsistent with the definition of a vegetable tannin proposed by White, but it seems probable that their historical association with the name tannin will continue. Like Tara tannin they are distinguished by the fact that their structures are based on carbohydrate fragments other than D-glucose. Hamameli tannin was first isolated by Von Grüttner[61] in 1898 from the bark of the witch hazel *Hamamelis virginica* and the work of Freudenberg[22] showed it to possess the structure of a digalloyl hexose; the structure of the sugar unit of the tannin being shown to be that of α-oxymethyl-D-ribose (45) —hamamelose—by Schmidt[62] in later work. Methylation and hydrolysis gave only 3,4,5-tri-O-methylgallic acid (24) indicating the absence in the compound of depside linkages. The tannin formed a methyl acetal (showing the aldehydic group to be unsubstituted) which when hydrolysed with alkali yielded methyl hamameloside (46) thus favouring the furanose formulation for the hexose molecule in the tannin structure. The location of the two galloyl groups on the sugar residue has not clearly been proved but the available evidence points to (47), in which both primary alcoholic groups are esterified, as the most probable structure. Schmidt has put forward[63] on this basis the interesting proposition that the tannin may be formed in the plant from two molecules of 3-O-galloylglyceraldehyde or one of this substance and one of O-galloyldihydroxyacetone. The tannin may have a wider natural distribution than has been anticipated and in recent work it was isolated from the bark of *Castanea sativa* by Mayer and Kunz[64].

The crystalline Acer tannin was obtained from the dried leaves of the Korean maple (*Acer ginnale*) by Perkin and Uyeda[21]. On prolonged acid hydrolysis it gave 1,5-anhydro-D-glucitol (48) and two molecules of gallic acid and Kutani[65] has presented evidence in favour of the 3,6-di-*O*-galloyl formulation (49) for this compound. Thus diazomethane methylation gave a hexamethylether which did not react with periodic acid but formed a di-acetate and on hydrolysis gave only 3,4,5-tri-*O*-methylgallic acid (24). The tannin Kutani reasoned therefore contained two single galloyl ester groups and two non vicinal hydroxyl groups. The 3,6-di-*O*-galloyl structure (49) was favoured[65] on the basis of the absence of reactions typical of a primary alcoholic group in both the tannin and its hexamethyl ether.

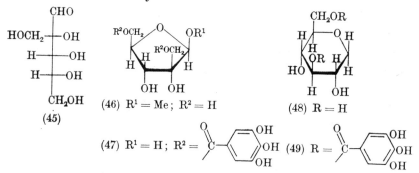

$$(45)$$

$$(46) \ R^1 = Me; \ R^2 = H$$

$$(47) \ R^1 = H; \ R^2 = \text{galloyl}$$

$$(48) \ R = H$$

$$(49) \ R = \text{galloyl}$$

4. MINOR POLYPHENOLS OF HYDROLYSABLE TANNIN EXTRACTS

Perhaps because of the difficulties of isolation and characterization and also of their assumed unimportance in the tanning action the distribution of the simpler polyphenols in vegetable tannin extracts has not been subject to a detailed study. Paper chromatographic analysis has shown several polyphenols to be widely distributed, in many cases these may represent breakdown products of the tannins themselves or they may be intermediates in the biosynthesis of these substances. Whichever is the case some knowledge of the structure and properties of these low molecular weight polyphenols is clearly desirable in order to describe more fully the nature of the tannin extracts.

β-D-*glucogallin*

The crystalline β-D-glucogallin (50) (β-1-*O*-galloyl-D-glucose) which was first isolated from the roots of Chinese rhubarb[42] (*Rheum officinale*) has been identified paper chromatographically in other extracts (Tara, Myrobalans, Eucalyptus)[2, 66, 67] and is also a product of incomplete breakdown of several gallotannins. Its structure was confirmed by a synthesis by Fischer and Bergmann[51] in which β-1-(tri-*O*-acetylgalloyl)-2,3,4,6-tetra-*O*-acetyl-D-glucose (51), prepared by condensation of tri-*O*-acetylgalloyl chloride and β-2,3,4,6-tetra-*O*-acetyl-D-glucose (52), was deacetylated with ammonia. Fischer and Bergmann also obtained heptaacetyl-α-glucogallin (53) by a similar condensation of α-2,3,4,6-tetra-*O*-acetyl-D-glucose (54) and tri-*O*-acetylgalloyl chloride, but were unable to obtain a crystalline product on de-acetylation. In a re-investigation of the reaction Schmidt and Herok[68] obtained a strongly dextrorotary crystalline compound which, however, exhibited mutarotation. This compound Schmidt and Reuss[69] formulated as α-2-*O*-galloyl-D-glucose (55) and suggested that it arose from

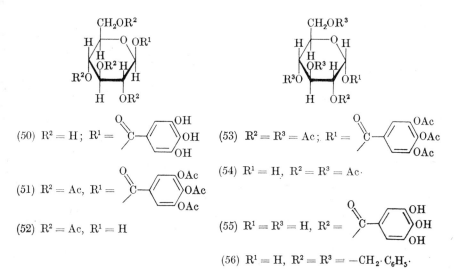

(50) $R^2 = H$; $R^1 =$

(51) $R^2 = Ac$, $R^1 =$

(52) $R^2 = Ac$, $R^1 = H$

(53) $R^2 = R^3 = Ac$; $R^1 =$

(54) $R^1 = H$, $R^2 = R^3 = Ac$·

(55) $R^1 = R^3 = H$, $R^2 =$

(56) $R^1 = H$, $R^2 = R^3 = -CH_2 \cdot C_6 H_5$·

(53) by migration of the galloyl group (1 → 2) during the de-acetylation; the rearrangement being favoured in this case by the *cis* orientation of the hydroxyl groups on C_1 and C_2[70]. Schmidt and Schmadel[71] later obtained independent syntheses of both α- and β-D-glucogallins by condensation of tri-*O*-benzylgalloyl chloride and α-2,3,4,6-tetra-*O*-benzyl-D-glucose (56), separation of the isomeric α- and β-heptabenzyl derivatives followed by hydrogenation to give the required products. Schmidt and his colleagues have also described in a series of papers the synthesis and properties of several other mono-*O*-galloyl-D-glucose derivatives (2-α-, 2-β-,[69] 3-[72] and 6-[72] esters) but none of these has been recorded as a natural product.

3,6-*Di-O-galloyl-* and β-1,3,6-*tri-O-galloyl*-D-glucose

From the alcoholic extracts of Myrobalans (*Terminalia chebula*) a di-*O*-galloyl and tri-*O*-galloyl-D-glucose derivative have been obtained after counter-current distribution[45]. Both may also be isolated[73, 74] as products of partial hydrolysis of chebulinic acid, a principal constituent of the tannin extract. The structure of the di-*O*-galloyl glucose derivative, deduced from observations of the periodate consumption of the corresponding hexamethyl methyl-glucoside, was confirmed as that of 3,6-di-*O*-galloyl-D-glucose (57) by synthesis[72] (Fig. 6).

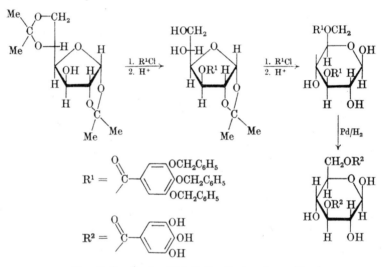

FIG. 6. Synthesis of 3,6-di-*O*-galloyl-D-glucose[72].

Partial hydrolysis of chebulinic acid at 60° yielded a non reducing tri-*O*-galloylglucose derivative identical with the tri-*O*-galloyl ester isolated from Myrobalans extract. Further hydrolysis of this compound yielded 3,6-di-*O*-galloyl-D-glucose (57) and hence the tri-*O*-galloyl ester must be formulated as the 1,3,6-tri-*O*-galloyl derivative (62) which was confirmed by synthesis[75] (Fig. 7). Condensation of tri-*O*-benzylgalloyl chloride with 2,4-di-*O*-benzyl-laevoglucosan (58) gave the ester (59) which was converted into the corresponding glucose derivative (60) by successive treatments with trifluroacetic anhydride and methanol. Further condensation with tri-*O*-benzylgalloyl chloride and hydrogenation of the tri- ester (61) gave the required galloyl derivative (62). The β-configuration of the C-1 ester linkage in both the synthetic and natural products was shown by anomerization of the fully acetylated derivative to the corresponding α-form (with concomittant increase in the dextro-rotation) by the action of boron trifluoride in chloroform.

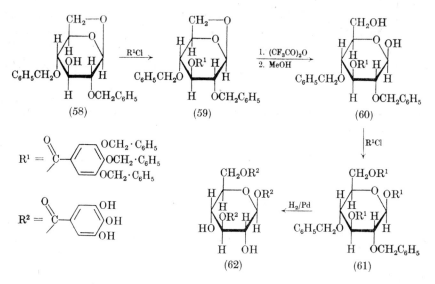

FIG. 7. Synthesis of β-1,3,6-tri-*O*-galloyl-D-glucose[75].

Other galloyl esters

Black tea is manufactured from the terminal shoots of the tea bush (*Camellia sinensis*) and derives its characteristic colour and flavour in part from oxidation products of polyphenols present in the green leaf. At least three of these are galloyl esters of a distinctive type—the 3-galloyl esters of (−)-epigallocatechin (19) and of (−)-epicatechin (18) (both isolated by Tsujimura[44] and later by other workers[76, 77]) and theogallin provisionally regarded as the 3-*O*-galloyl ester of quinic acid[43] (65). Tsujimura[44] isolated (−)-epicatechin gallate (18) during his studies on tea tannins and believed the galloyl ester to be located at the 3-position, but attempts to synthesize a compound of this type have produced only amorphous materials. Freudenberg[78] succeeded in making the 3-gallate of (+)-catechin and characterized the compound as its crystalline acetate. However, this gallate has not been reported as a natural product.

Theogallin, an amorphous acidic polyphenol, was isolated by Roberts from green tea and on the basis of analytical data and the biogenetic analogy to chlorogenic acid it was assigned the 3-*O*-galloylquinic acid structure[43] (63). A compound of this structure was synthesized by

Haslam, Haworth and Lawton[60] and showed many similarities to theogallin but lack of sufficient descriptive data on the natural product has prevented a more complete comparison.

Tannic acid as isolated from Chinese galls has a pronounced acidic reaction which is due[28] principally to the presence of three acidic components in the extract—gallic (1), m-digallic (14) and m-trigallic (15) acids. These acids may also be isolated and identified as products of the mild acid hydrolysis of Chinese and Sumach gallotannins. The structures of (14) and (15) have been inferred as linear m-depsides from degradative work on the tannins and later proved by synthesis. Fischer, Bergmann and Lipschitz[79] synthesized m-digallic acid by condensation of 3,5-di-O-acetylgallic acid (64) and 3,4,5-tri-O-acetylgalloyl chloride (65)

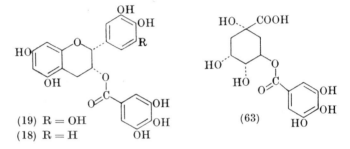

(19) R = OH
(18) R = H

(63)

to give the depside (66) which upon deacetylation rearranged to the meta depside (14). This method has several drawbacks notably the attendant possibilities of hydrolysis of the depside linkage during the deacetylation but White and King[28] successfully applied the same principle in a synthesis of the related m-trigallic acid (15). Unequivocal syntheses of both m-digallic and m-trigallic acids have been accomplished by Crabtree, Haslam, Haworth and Mills[80] and employed protecting groups which could be readily removed by hydrogenation. The route used for the preparation of m-trigallic acid is shown along with the original method of Fischer for m-digallic acid in Fig. 8.

Derivatives of the hydroxycinnamic acids

Widely occurring in the plant kingdom are esters and glycosides of the hydroxycinnamic acids[81] (67 a–d) and their presence in certain hydrolysable tannin extracts has been affirmed. As esters the acids are normally associated with glucose or quinic acid[81] (68; all R's = H) but esters with shikimic acid,[82] tartaric acid[83] and with anthocyanins[84] have also been reported. Of the esters with quinic acid the most commonly occurring and most extensively studied is the 3-O-caffeoyl derivative[85] (68; R¹ = R² = R⁴ = R⁵ = H, R³ = 3,4-dihydroxycinnamoyl). Since 1950 the isolation of at least four other caffeoyl esters of quinic acid has been claimed. The structure of cynarin [86] from the artichoke has been shown to be 1,4-di-O-caffeoylquinic acid (68; $R^1 = R^4 = 3,4$-dihydroxycinnamoyl, $R^2 = R^3 = R^5 = H$) and that of neochlorogenic acid [5, 87] (from peaches, prunes, tobacco and other

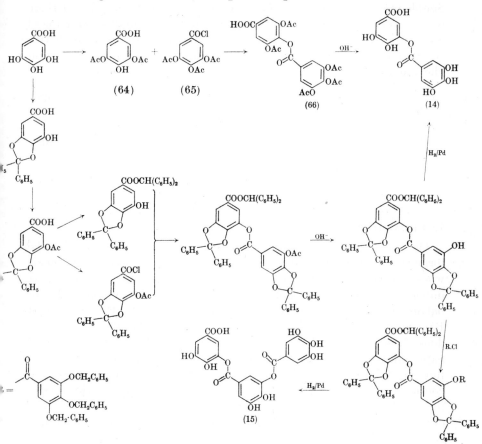

FIG. 8. Synthesis of m-digallic and m-trigallic acids [79, 80].

sources) as 5-O-caffeoylquinic acid (68; $R^1 = R^2 = R^3 = R^4 = H$, $R^5 = 3,4$-dihydroxycinnamoyl). No structures have been proven for isochlorogenic acid [88] (coffee) and pseudochlorogenic acid [89] (sweet potato) although the former is in all probability a mixture of di-O-caffeoylquinic acids. Esters of the remaining hydroxycinnamic acids although shown to be ubiquitous by paper chromatographic analysis [81] do not appear to be present in plant tissues in such high concentration as their caffeoyl counterpart and it could be concluded from this that this latter acid is an end product of the plant's metabolism.

Esters of gallic acid and the hydroxycinnamic acids (67a–d) form the two major groups of naturally occurring phenolic esters and it is of interest to comment briefly on their predominant characteristics. Simple mono- and di-esters of both groups are known but as yet there is no authenticated record of the isolation of a polyhydroxycinnamoyl ester from Nature and thus the polygalloyl structures associated with the gallotannins are seen clearly to be distinctive amongst natural phenolic esters.

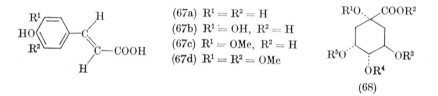

$$(67a)\ R^1 = R^2 = H$$
$$(67b)\ R^1 = OH,\ R^2 = H$$
$$(67c)\ R^1 = OMe,\ R^2 = H$$
$$(67d)\ R^1 = R^2 = OMe$$

(68)

VIII. ELLAGITANNINS

1. ELLAGIC ACID

The ellagitannins are differentiated from the gallotannins by the formation on acid hydrolysis of ellagic acid (2) and occasionally acids whose structures may be related biogenetically to ellagic acid. The most important sources of ellagitannins[63] are (Table I)—Valonea, Myrobalans, Divi-divi and Algarobilla—but, according to the surveys of Bate-Smith[90], ellagic acid itself occurs in about 75 families of dicotelydons most consistently in the orders Fagales, Myrtiflorae, Rosales, Sapindales and Geraniales. Ellagic acid is also found frequently in conjunction with esters of gallic acid which is perhaps not too surprising considering the ease with which such esters readily form this acid[91] (e.g. photo oxidation,[92] autoxidation in alkaline media[93]). The presence of ellagic acid in an extract is not necessarily therefore a reliable indication of the presence of ellagitannins. Ellagic acid which has a very low solubility in water and normally may be crystallized from aqueous pyridine as pale yellow needles, is conveniently identified paper chromatographically by its distinctive R_f values and light blue fluorescence under ultra-violet light (Tables II and III). Two partially methylated derivatives of ellagic acid have also been isolated from Nature; the 3,3'-dimethylether (69) obtained from the roots of *Euphorbia formosana*[94] and the 3,3',4'-trimethyl ether (70) from the bark of *Eugenia maire*.[95] Both of these structures have been confirmed by synthesis[96, 97] and these are shown in Fig. 9 along with other reactions[98, 99] of ellagic acid.

Schmidt, whose contributions to ellagitannin chemistry have been outstanding, has shown[100] that ellagic acid as such does not occur in the structure of the tannins but that it is produced by lactonization of

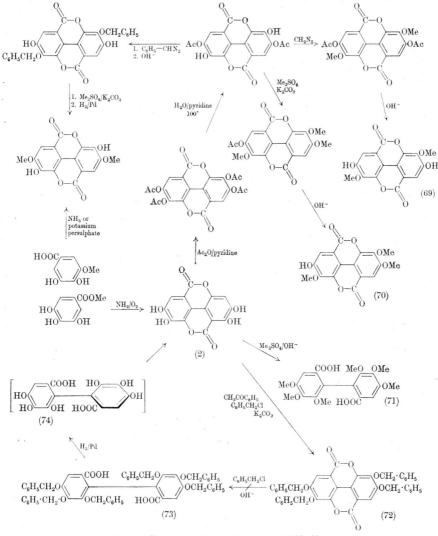

FIG. 9. Some reactions of ellagic acid[96-99].

hexahydroxydiphenic acid (74) formed by hydrolytic cleavage of the hexahydroxydiphenoyl group (76) from the ellagitannin structures. Thus although acid hydrolysis of corilagin, chebulagic acid and other

ellagitannins gave ellagic acid, hydrolysis of the corresponding diazomethane methylated tannins yielded instead hexamethoxydiphenic
acid (71) usually in one of its optically active forms. Schmidt has found
hexahydroxydiphenic acid (74) to occur in Divi-divi and Myrobalans[101]
in the dextro rotary form and in Algarobilla[102], Valonea[103] and
Knoppern nuts[63] in the laevorotary form and considers this good
evidence in favour of the supposition[104] that the hexahydroxydiphenoyl
group (76) in ellagitannins arises by oxidative coupling of two suitably
disposed galloyl groups (75) in a galloylated sugar rather than the
alternative of esterification of the sugar by a preformed molecule of
hexahydroxydiphenic acid.

The optical activity of the natural hexahydroxydiphenic acids is due
to restricted rotation about the diphenyl linkage and both the hexabenzyl[99] (73) and hexamethyl[98] (71) derivatives prepared from ellagic
acid have been resolved into their optically active forms. Benzyl
chloride reacts unusually with ellagic acid in aqueous alkaline solution
to give a deep red pigment ellagorubin for which two structures have
been proposed by Schmidt and Jurd respectively. Both of these structures were postulated as being derived by processes involving C-
benzylation of the aromatic nuclei. Benzylation in acetophenone with
potassium carbonate followed a more normal course[99] with formation
of the tetrabenzyl ether, which was also produced[99] by the action of
phenyldiazomethane. Further treatment of the tetrabenzyl ether with
benzyl chloride and alkali gave the hexabenzyl derivative (73) which
Schmidt and his colleagues[99] were able to resolve into its optically
active forms via its quinidine and cinchonine salts. The optically active
forms of the benzyl ether (73) were not readily racemized in contrast to
those of the corresponding hexamethoxy derivatives, prepared in similar
ways, which had a half life of 14 h 45 min in boiling alkali. Reduction
of the isomeric hexabenzyl ethers (73) gave the corresponding hexahydroxydiphenic acids (74) but these were not isolable since at 20° in
methanol they had a half life of racemization of 4 hr 40 min. However,
reduction of the ($-$)-hexabenzyl ether followed by immediate treatment of the solution with diazomethane gave ($+$)-dimethyl hexa-
methoxydiphenate identical with the dimethyl ester of the hydrolysis
product of methylated corilagin[105] or chebulagic acid[101].

Four substances found in ellagitannins or ellagitannin extracts which
bear a close structural and biogenetic relationship to ellagic acid are
chebulic acid[106] (83), brevifolin carboxylic acid[25] (84), valoneaic acid
dilactone[107] (82) and dehydrodigallic acid[108] (81). As with ellagic acid
so also with the first three of these substances (82, 83, 84) it is considered that each one is produced by cleavage of a particular fragment

(78, 79, 80 respectively) from an ellagitannin followed by lactonization. Dehydrodigallic acid (81) is isolated in the form in which it presumably occurs in the tannin. Schmidt[63, 104] has indicated how each of these

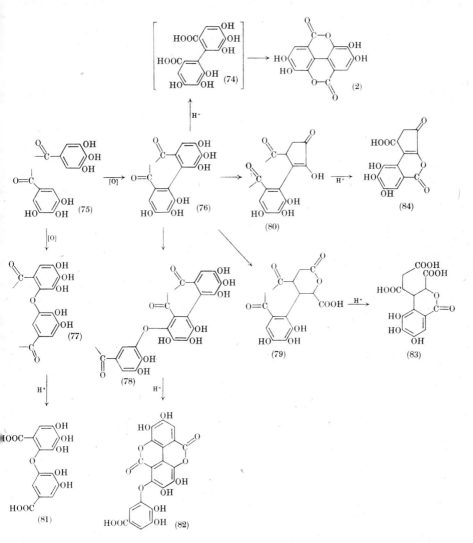

FIG. 10. Biogenetic relationships in the ellagitannins[68, 104].

substances (81, 82, 83, 84) may be derived by oxidation, reduction and ring cleavage of a hexahydroxydiphenoyl precursor and his arguments are considered in more detail later but are summarized in Fig. 10.

2. CORILAGIN

Corilagin was the first ellagitannin to be isolated[24] and character-ized[105]. It is a crystalline substance and has been isolated from Divi-divi[24] and Myrobalans[102] and identified in a provisional way by paper

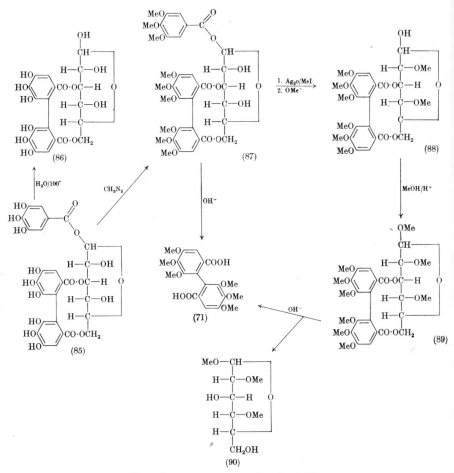

FIG. 11. Some reactions of corilagin[105].

chromatographic methods in other plant extracts. Acid hydrolysis of corilagin gave equimolecular proportions of glucose, gallic and ellagic acids and controlled hydrolysis with water at 100° produced gallic acid and an amorphous substance (86) which in contrast to corilagin gave a positive reaction with anisidine hydrogen phthalate; indicating therefore that the readily hydrolysed galloyl ester group was attached originally to the glucosidic hydroxyl group of corilagin. Anomerization of the fully acetylated corilagin with boron trifluoride according to the procedure of Lindberg[57] showed the glucosidic galloyl group to have the β-configuration. Nonamethyl corilagin (87) resulted from diazomethane methylation of corilagin and this on hydrolysis gave 3,4,5-tri-O-methylgallic acid and (+)-hexamethoxydiphenic acid (71). Further methylation of nonamethyl corilagin with silver oxide and methyl iodide gave an undecamethyl ether which treated with a limited amount of potassium methoxide gave an octamethyl ether of hexahydroxydiphenoyl-D-glucose (88). Treatment with methanolic hydrogen chloride converted the latter into a mixture of α- and β-methylglucosides (89) which when hydrolysed with alkali gave (+)-hexamethoxydiphenic acid (71) and a mixture of glucosides from which β-methyl-2,4-di-O-methyl-D-glucoside (90) was obtained in a crystalline form. Thus the attachment of the (+)-hexahydroxydiphenoyl residue to the 3,6-positions on the glucose molecule was established and the structure of corilagin followed as (85), Fig. 11 (the structures of the ellagitannins are most conveniently represented using the Fischer formulation for glucose as opposed to the Haworth projections used elsewhere in this text). A synthesis of corilagin has not been reported.

3. CHEBULIC, CHEBULINIC AND CHEBULAGIC ACIDS

Chebulinic[23] and chebulagic acids[26] are two crystalline tannins which are isolated from the extract of the fruit of *Terminalia chebula*, commonly known as Myrobalans. Both compounds liberate chebulic acid (83) on hydrolysis[109] and therefore bear a close structural relationship one to another. However, although it is convenient to consider for this reason the chemistry of chebulinic acid alongside that of chebulagic acid the former does not liberate ellagic acid on hydrolysis and it is also necessary therefore to point out that it is not strictly an ellagitannin.

Chebulic acid[73] (formerly known as "split acid") is an amorphous optically active trihydroxy–tricarboxylic acid $C_{14}H_{12}O_{11}$ and its structure (83) was deduced by Schmidt and Mayer[106, 109, 110] and confirmed by Haworth and his collaborators[23, 111]. The acid had the properties of a lactone and with diazomethane gave an amorphous hexamethyl derivative which on hydrolysis gave trimethylchebulic acid (91). Oxidation of the acid or its tri-O-methyl ether gave an optically active tetracarboxylic acid, isolated as its lactone $C_8H_8O_8$ for which structure (92) was advanced and confirmed by synthesis[63] (Fig. 12). The triamide of trimethylchebulic acid gave one mole of sodium cyanate when

6*

treated with alkaline hypochlorite, indicating the presence of a potential α-hydroxy acid system in chebulic acid.

Oxidation of trimethylchebulic acid with potassium ferricyanide gave 3,4,5-trimethoxyphthalic acid and pyrolysis of the same material gave succinic acid and the isocoumarin carboxylic acid[23] (93) whose structure Haworth, Pindred and Jefferies[111] confirmed by synthesis

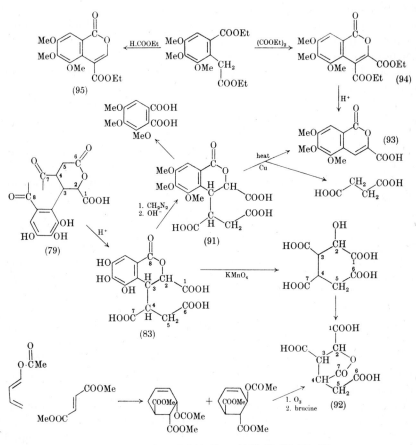

FIG. 12. Degradation of chebulic acid [23, 63, 106, 109-111].

(Fig. 12). The constitution assigned to (93) resulting from hydrolysis and decarboxylation of the intermediate (94) in the synthesis was proved to be correct by preparation of its ethyl ester which differed from that of the isomeric product (95) prepared by the unequivocal route as shown. As a result of these various degradations (Fig. 12) the structure of chebulic acid followed most satisfactorily as (83). This structure has not been confirmed by any rational synthesis.

Comparison of the molecular formulae of chebulic and ellagic acid shows that the former contains three more molecules of water and Schmidt was led to suggest[63] on this basis that ellagic acid might give

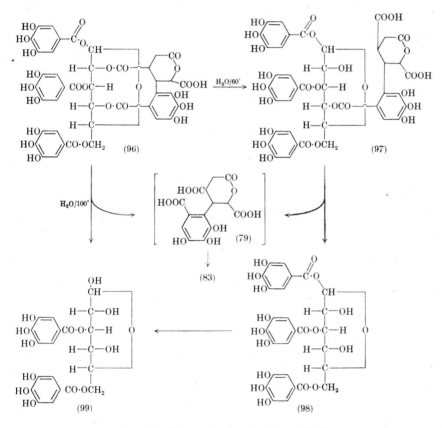

FIG. 13. Hydrolysis of chebulinic acid[73, 74].

rise to chebulic acid by a series of redox reactions. Since ellagic acid has been shown to be derived as a result of cleavage of the hexahydroxydiphenoyl group from tannin molecules[100] Schmidt argued that it was reasonable to assume that the redox reactions postulated for the ellagic acid → chebulic acid change should take place with the hexahydroxydiphenoyl group attached to the sugar. These considerations led[104] to the alternative formulation of chebulic acid as (79) when combined in natural tannin molecules such as chebulinic and chebulagic acids.

Chebulinic acid (96), the major constituent of the Myrobalans extract, was the first hydrolysable tannin to be obtained in a crystalline form[112]. Freudenberg[73] showed that on complete hydrolysis it gave one mole of glucose, three of gallic and one of chebulic acid and on partial hydrolysis at 100° with water it gave a di-O-galloyl-D-glucose derivative. Schmidt[72, 113] confirmed the structure of the latter by synthesis as 3,6-di-O-galloyl-D-glucose (99) (see p. 118) and by the action of water at 60° was able to isolate[74] further intermediates in the hydrolysis of chebulinic acid. Hydrolysis for 7–8 days at 60° gave a tri-O-galloyl-D-glucose whose structure was shown to be that of β-1,3,6-tri-O-galloyl-D-glucose (98) by analysis and synthesis[75] (see

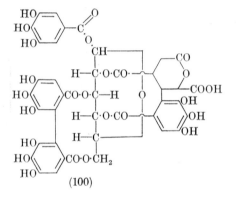

(100)

p. 118). Neochebulinic acid (97), isolated[75] after treatment with water at 60° for 2–4 days, was a crystalline dicarboxylic acid and on complete methylation and hydrolysis with potassium methoxide gave 2-O-methyl-D-glucose. Since both chebulinic and neochebulinic acids showed no mutarotation in solution (i.e. both contained a stable lactone ring) it was deduced that the conversion of chebulinic → neochebulinic acid involved cleavage of an aliphatic ester bond attached to glucose at the 2- position. On the basis of this evidence (Fig. 12), and the biogenetic considerations outlined above which lead to the correct formulation of chebulic acid in the bound form[104] (79), the structure (96) was put forward for chebulinic acid[74].

The crystalline chebulagic acid (100) occurs in Myrobalans and Dividivi and it bears a close structural relationship to both corilagin (85) and chebulinic acid (96). On vigorous acid hydrolysis[26, 114] it gave equimolecular proportions of glucose, gallic, ellagic and chebulic acids.

Hydrolysis of the tridecamethyl derivative following prolonged treatment with diazomethane gave (+)-hexamethoxydiphenic acid[101] (71). In water at 60° hydrolysis proceeded in a stepwise manner in much the same way as indicated for chebulinic acid and by following the changes either polarimetrically or chromatographically Schmidt was able to isolate various intermediates in the reaction[115]. After several days in water at 60° corilagin (85) and chebulic acid (83) were formed and interruption of the reaction at an earlier stage (3 days) allowed the isolation of neochebulagic acid which on methylation and hydrolysis gave 2-O-methyl-D-glucose[116]. The change from chebulagic to neochebulagic acid was thought to involve the aliphatic carboxyl group of the chebulic acid molecule in an analogous manner to the chebulinic → neochebulinic acid transformation. With this information in mind the structure of chebulagic acid followed as (100)[116].

4. BREVILAGIN I, II AND TERCHEBIN

Three neutral tannins isolated and described recently by Schmidt and his collaborators[45, 102, 117] possess structural units which form important links in Schmidt's theoretical schemes for the biogenesis of the ellagitannins[104]. Brevilagin I and II, isolated from algarobilla (fruit pods of *Caesalpinia brevifolia*) in 1–2 and 0·5% yields respectively contain the dehydrohexahydroxydiphenoyl residue (101) in their structures and terchebin (102), from the Myrobalans extract (1·8% yield), possesses the very closely related structural residue of isohexahydroxydiphenic acid. Brevilagin I, a yellow crystalline substance, gave a tetrakis-phenylhydrazone and although tests for free and combined ellagic acid were negative, acid hydrolysis did give varying amounts of this acid in admixture with other products including D-glucose. Hydrolysis with concentrated hydrochloric acid was rapid (10 min) and gave the previously unknown chlorellagic acid (103). Brevilagin I reacted with two moles of o-phenylenediamine to give an amorphous yellow product which on acid hydrolysis yielded the crystalline salmon-red phenazine whose structure (105) was deduced by standard procedures. Schmidt considered that the presence of two dehydrohexahydroxydiphenoyl residues (101), probably in hydrated form, in brevilagin I are responsible for the unusual products resulting from acid hydrolysis and for the formation of the characteristic phenazine on reaction with o-phenylenediamine. The chlorellagic acid (103) it was suggested was formed by addition of hydrogen chloride to the quinonoid system of (101) followed by acid hydrolysis and subsequent lactonization (Fig. 14).

Brevilagin II, an amorphous yellow substance, gave a positive test for bonded ellagic acid and methylation with diazomethane gave a nonamethyl derivative which on alkaline hydrolysis gave (−)-hexa-methoxydiphenic acid (71), thus confirming the presence of a hexa-hydroxydiphenoyl group in the tannin. Concentrated hydrochloric acid

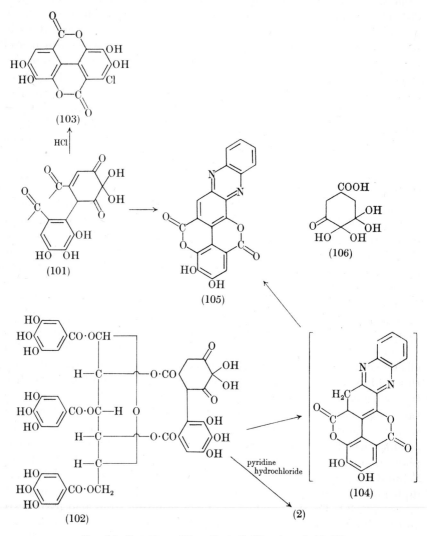

Fig. 14. Reactions of brevilagin I, II and terchebin[106].

hydrolysis of brevilagin II led to a mixture of ellagic and the chlor-ellagic acid (103) and the presence of the same dehydrohexahydroxydi-phenoyl unit (101) as in brevilagin I was confirmed by the preparation of the phenazine (105) in high yield following reaction of the tannin with one mole of *o*-phenylenediamine. Although the component parts of the structures of both brevilagin I and II have thus been identified the location of the positions of esterification of the various residues on the glucose nucleus of the tannins has not been defined.

The third substance from Myrobalans, terchebin (102), was again yellow and crystalline and with a mole of *o*-phenylenediamine it gave a brilliant yellow condensation product. The latter on treatment with acetic acid gave the known β-1,3,6-tri-*O*-galloyl-D-glucose (98) and the previously described phenazine (105). Hydrolysis of terchebin with hydrochloric acid did not give either chlorellagic or ellagic acids but if terchebin was first warmed with pyridine hydrochloride prior to acid hydrolysis ellagic acid was formed. Schmidt has interpreted this sequence of reactions in terms of the structure (102) for terchebin containing the isohexahydroxydiphenoyl group. Thus formation of the phenazine (105) may be rationalized in terms of the loss of two hydrogen atoms from the intermediate (104) and the isomerization of terchebin with pyridine hydrochloride is analogous to the known isomerization of (106) into gallic acid[120]. The production of the isohexahydroxydiphenoyl unit from that of hexahydroxydiphenoyl has been postulated by Schmidt as one of the key steps in the biosynthesis of many of the unusual groups found in the ellagitannins. Subsequent reactions involving ring fission and decarboxylation to give the chebulic acid (79) and brevifolin carboxylic acid (80) units are then suggested as occurring with the isohexahydroxydiphenoyl unit. Support for these ideas comes from the observation that chebulic acid (83) was formed on treatment of terchebin with base and that brevifolin carboxylic acid (84) was formed by heating terchebin in water. It is perhaps surprising in view of these facile changes undergone by the isohexahydroxydiphenoyl group in terchebin that it has been found possible to isolate this particular compound from tannin extracts.

5. BREVIFOLIN AND BREVIFOLIN CARBOXYLIC ACID

Algarobilla is a tanning material obtained from Middle and South America and consists principally of the ground fruit pods of plants of the *Acacia* species, usually *Caesalpinia brevifolia*. Two of the first compounds which were isolated[25] and later characterized from algarobilla were brevifolin, $C_{12}H_8O_6$ (107) and brevifolin carboxylic acid, $C_{13}H_8O_8$ (84). Brevifolin (107) is merely the decarboxylation product of brevifolin carboxylic acid (84) and it is probable, following Schmidt's hypotheses[104], that both are derived in the tannin extract from a bonded form of the latter (80) which occurs in certain tannin molecules present

methyl ether (109), which was readily converted to the tri-O-methyl ether by treatment with acid, thus substantiating the presence of an enol-lactone ring in brevifolin. Brevifolin like the parent acid formed a 2,4-dinitrophenylhydrazone and the presence of a methylene group adjacent to the ketonic group was confirmed by the production with benzaldehyde of a benzylidene derivative. Oxidation of tri-O-methyl-brevifolin led to a variety of products, potassium ferricyanide gave 3,4,5-trimethoxyphthalic acid, potassium permaganate yielded succinic acid and alkaline hydrogen peroxide afforded the tricarboxylic acid (110). The latter observation indicated the presence of a potential α-diketone system in brevifolin which was accordingly assigned[25] the structure (107). Two independent syntheses[118, 119] of tri-O-methyl-brevifolin have confirmed this structure. Haworth and Grimshaw[119] oxidized flavellagic acid (111) with hydrogen peroxide to the acid (112); methylation and chain extension gave (113) which on treatment with phosphorus pentoxide yielded tri-O-methylbrevifolin (107) (Fig. 15).

Since brevifolin carboxylic acid was so readily decarboxylated it must have the structure of a β-keto acid or a vinylogous β-keto acid (84) and the latter was confirmed by a synthesis of its tri-O-methyl ether (114) by Schmidt and Eckert[121]. Coupling of diazotized methyl-2-amino-3,4,5-tri-O-methylgallate (115) and cyclopenta-3, 4-dione-1-carboxylic acid (116) in the presence of cupric acetate gave directly small yields of (114). Brevifolin carboxylic acid (84) as isolated from algarobilla is optically inactive. Since Schmidt has suggested that the acid is derived from hexahydroxydiphenic acid which occurs in both the ($+$) and ($-$) forms in various tannins, very probably brevifolin carboxylic acid is combined in an analogous optically active form and becomes racemized during or after its cleavage from the tannin[102].

6. VALONEAIC ACID DILACTONE AND DEHYDRODIGALLIC ACID

In an exactly parallel manner to brevifolin carboxylic acid, dehydro-digallic acid[122] (81) and valoneaic acid dilactone[123] (82) probably form parts of the structure of much larger tannin molecules in the particular plant tissues from which they have been isolated. Their formation can, according to the postulates of Schmidt[104], be envisaged as falling within the general scheme of oxidative coupling between galloyl groups in a gallotannin molecule. In the case of dehydrodigallic acid (81) coupling is envisaged as occurring with the formation of a diphenyl ether linkage between two galloyl groups (77) (Fig. 10) and in the case of valoneaic acid dilactone (82) of an ether linkage between a galloyl and a hexahydroxydiphenoyl group (78) (Fig. 10).

Dehydrodigallic acid (81) was isolated by Mayer[122] from the leaves, young shoots and bark of the Spanish chestnut (*Castanea vesca*) and formed a penta-*O*-methyl ether which gave a dimethyl ester[122]. With dilute alkali the penta-*O*-methyl ether was recovered unchanged but dehydrodigallic acid itself underwent an unusual fission of the diphenyl ether linkage to give gallic acid in approximately 50% yield. A study of model compounds showed[124] that this ether cleavage only manifested

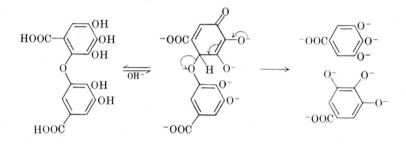

itself when one of the two phenyl groups contained 3 vicinal hydroxyl groups. Mayer, Fikentscher, Schmidt and Schmidt[124] have discussed a mechanism for this reaction in terms of the β-elimination process shown above.

Treatment of dehydrodigallic acid with sulphuric acid gave the xanthone (117) which on methylation gave a penta-*O*-methyl ether

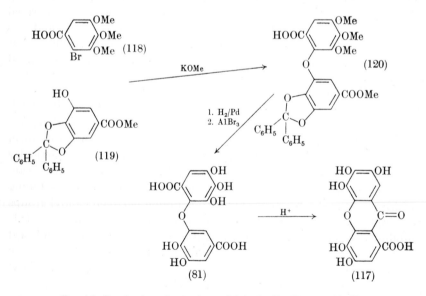

FIG. 16. Synthesis and reactions of dehydrodigallic acid[122, 125].

identical with that prepared similarly from penta-O-methyldehydrodigallic acid. On the basis of this evidence Mayer suggested[122] structure (81) for dehydrodigallic acid and this was confirmed in a synthesis[125] of the acid and its penta-O-methyl derivative (Fig. 16). Condensation of 2-bromo-3,4,5-tri-O-methyl gallic acid (118) and methyl 3,4-diphenyl-methylenedioxy-5-hydroxybenzoate (119) gave (120) which on hydrogenation and subsequent demethylation with aluminium bromide gave dehydrodigallic acid.

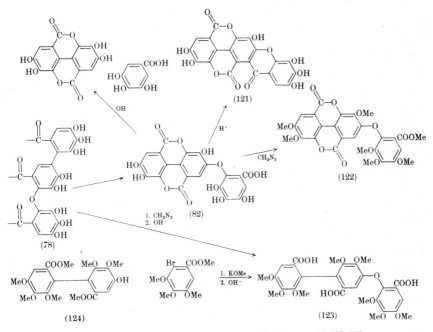

Fig. 17. Some reactions of valoneaic acid dilactone[96, 103, 123].

Hydrolysis of valonea tannin, obtained from the acetone extraction of the acorn cups of the bearded oak (*Quercus valonea*) yielded along with substantial quantities of ellagic acid an optically inactive crystalline phenolic acid, valoneaic acid dilactone[123] (82). Its structural similarity to dehydrodigallic acid (81) was borne out in its chemical reactions. With sulphuric acid valonea xanthone (121) was formed and with alkali an analogous splitting of the diphenyl ether link occurred, giving gallic and ellagic acid in rather less than 50% yield. Treatment of valoneaic acid dilactone with diazomethane gave a hexamethyl ether methyl ester (122) and further methylation with dimethyl sulphate and alkali gave the octamethyl derivative (123) giving proof of the presence of two lactone rings in the original compound. The octamethyl derivative (123) was isolated[103] in an optically active (−) form along with (−) hexamethoxydiphenic acid (71) following methylation of crude valonea with diazomethane and hydrolysis. The isolation of (123) in an optically active form proves that valoneaic acid dilactone is bound in tannin molecules in an entirely different form (78) to that which is isolated (82) and adds further weight to Schmidt's proposal[104] that the molecule is formed by oxidative coupling of three galloyl groups in a gallotannin molecule (Fig. 10).

On the basis of this evidence and its general chemical similarity to dehydrodigallic acid (81) valoneaic acid dilactone was assigned the structure (82) and the general features of this structure have been confirmed in a synthesis[96] of the octamethyl derivative (123). Condensation of the hexahydroxydiphenic acid derivative (124) with 2-bromo-3,4,5-tri-O-methylgallic acid (118) and hydrolysis of the product with alkali gave the required product (123) (Fig. 17).

7. MISCELLANEOUS ELLAGITANNINS

The surveys of plant species by Bate-Smith[90] indicated that ellagic acid (and therefore perhaps ellagitannins) was widely distributed in Nature. Ellagitannins have been discerned in the cambium and woody tissues of *Eucalyptus* species[67], in the leaves of certain *Acer* species[20] and juglanin, an amorphous tannin thought to be isomeric with corilagin, has been obtained from walnut skins (*Juglans regia*)[126]. The constitution of juglanin and these other tannins have not however been elucidated and there is at present considerable scope for further work in this field.

REFERENCES

1. Haworth, R. D., and Haslam, E., "Progress in Organic Chemistry", (W. Carruthers and J. W. Cook, eds.), Vol. 6, p. 1, Butterworths, London (1964).
2. Armitage, R., Bayliss, G. S., Gramshaw, J. W., Haslam, E., Haworth, R. D., Jones, K., Rogers, H. J., and Searle, T., *J. chem. Soc.*, 1842 (1961).
3. White, T., Kirby, K. S., and Knowles, E., *J. Soc. Leath. Trades Chem.*, **36**, 148 (1952).
4. Haslam, E., Haworth, R. D., Mills, S. D., Rogers, H. J., Armitage, R., and Searle, T., *J. chem. Soc.*, 1836 (1961).
5. Haslam, E., Makinson, K. G., Naumann, M. O., and Cunningham, J., *J. chem. Soc.*, 2137 (1964).
6. Bate-Smith, E. C., "Wood Extractives", (W. E. Hillis, ed.), p. 133, Academic Press, London (1962).

7. White, T., Kirby, K. S., and Knowles, E., *J. Soc. Leath. Trades Chem.*, **35**, 338 (1951).

8. Lindstedt, G., *Acta chem. scand.*, **4**, 448 (1940).

9. Haslam, E., unpublished observations.

10. Gibbs, H. D., *J. biol. Chem.*, **72**, 649 (1927).

11. King, F. E., King, T. J., and Manning, L. C., *J. chem. Soc.*, 563 (1957).

12. Smit, C. J. B., Joslyn, M. A., and Lukton, A., *Analyt. Chem.*, **27**, 1159 (1955).

13. Pro, M. J., *J. Ass. off. agric. Chem.*, **35**, 255 (1952).

14. Bray, H. G., Humphris, B. G., Thorpe, W. V., White, K., and Wood, P. B., *Biochem. J.*, **52**, 416 (1952).

15. Kursanov, A. L., and Zaprometov, M. N., *Biokhimiya*, **14**, 467 (1949).

16. Aulin-Erdtman, G., *Svensk Papp-Tidn.*, **56**, 287 (1953).

17. Swain, T., and Hillis, W. E., *J. Sci. Fd Agric.*, **10**, 63 (1959).

18. Knudson, L., *J. biol. Chem.*, **14**, 159 (1913).

19. Haslam, E., Haworth, R. D., Jones, K., and Rogers, H. J., *J. chem. Soc.*, 1820 (1961).

20. Haslam, E., unpublished results.

21. Perkin, A. G., and Uyeda, Y., *J. chem. Soc.*, **121**, 66 (1922).

22. Freudenberg, K., *Ber. dt. chem. Ges.*, **52**, 177 (1919).

23. Haworth, R. D., and de Silva, J. B., *J. chem. Soc.*, 3611 (1954).

24. Schmidt, O. T., and Lademann, R., *Justus Liebigs Annln Chem.*, **571**, 41, 232 (1951).

25. Schmidt, O. T., and Bernauer, K., *Justus Liebigs Annln Chem.*, **588**, 211 (1954).

26. Schmidt, O. T., and Lademann, R., *Justus Liebigs Annln Chem.*, **569**, 149 (1950).

27. Haslam, E., Haworth, R. D., and Keen, P. C., *J. chem. Soc.*, 3814 (1962).

28. White, T., and King, H. G. C., *J. chem. Soc.*, 3231 (1961).

29. Freudenberg, K., and Vollbrecht, E., *Hoppe-Seyler's Z. physiol. Chem.*, **116**, 277 (1921).

30. Dyckerhoff, H., and Armbruster, R., *Hoppe-Seyler's Z. physiol. Chem.*, **219**, 38 (1933).

31. Toth, G., and Barsony, G., *Enzymologia*, **11**, 19 (1943–5).

32. Toth, G., *Magy. Timar*, **5**, 1 (1944).

33. Dhar, S. C., and Bose, S. M., *Bull. Cent. Leath. Res. Inst.* (1964).

34. Madharakrishna, W., and Bose, S. M., *Bull. Cent. Leath. Res. Inst., Madras*, **8**, 153 (1961).

35. Park, C. R., and Johnson, L. H., *J. biol. Chem.*, **181**, 150 (1949).

36. White, T., "The Chemistry of the Vegetable Tannins", p. 13, Society of Leather Trades Chemists, Croydon (1956).

37. Grassmann, W., Stiefenhofer, G., and Endres, H., *Chem. Ber.*, **89**, 454 (1956).

38. Haslam, E., Haworth, R. D., Gramshaw, J. W., and Searle, T., *J. chem. Soc.*, 2944 (1952).

39. Freudenberg, K., Blummel, F., and Frank, T., *Hoppe-Seyler's Z. physiol. Chem.*, **164**, 262 (1927).

40. Freudenberg, K., "Tannin, Cellulose and Lignin", Springer-Verlag, Berlin (1930).

41. Chmielewska, I., and Kasprzyk, Z., *Nature, Lond.*, **196**, 776 (1962).

42. Gilson, E., *C. r. hebd Séanc. Acad. Sci.*, *Paris*, **136**, 385 (1903).
43. Roberts, E. A. H., and Myers, M. J., *J. Sci. Fd Agric.*, **11**, 701 (1958).
44. Tsujimura, M., *Sci. Pap. Inst. phys. chem. Res.*, *Tokyo*, **26**, 186 (1935).
45. Schmidt, O. T., *Leder*, **14**, 40 (1963).
46. Reddy, N. K., Rajadwai, S., Sastry, K. N. S., and Nayudamma, Y.. *Aust. J. Chem.*, **17**, 238 (1964).
47. Russell, A., and Tebbens, W. G., *J. Am. chem. Soc.*, **64**, 2274 (1942).
48. Fischer, E., *Ber. dt. chem. Ges.*, **52**, 809 (1919).
49. Karrer, P., Salomon, H. R., and Payer, J., *Helv. chim. acta.*, **6**, 3 (1923).
50. Karrer, P., Widmer, R., and Staub, M., *Justus Liebigs Annln Chem.*, **433**, 288 (1923).
51. Fischer, E., and Bergmann, M., *Ber. dt. chem. Ges.*, **51**, 1760 (1918).
52. Feist, K., *Ber. dt. chem. Ges.*, **45**, 1493 (1912).
53. Fischer, E., and Freudenberg, K., *Ber. dt. chem. Ges.*, **47**, 2485 (1914).
54. Lowe, *Z. analyt. Chem.*, **12**, 128 (1873).
55. Armitage, R., Haslam, E., Haworth, R. D., and Searle, T., *J. chem. Soc.*, 3808 (1962).
56. White, T., and King, H. G. C., *Chemy Ind.*, 1683 (1958).
57. Lindberg, B., and Swan, B., *Acta. chem. scand.*, **14**, 1043 (1960).
58. Haslam, E., Haworth, R. D., and Stangroom, J. E., unpublished observations.
59. Horler, D. F., and Nursten, H. E., *J. chem. Soc.*, 3786 (1961).
60. Haslam, E., Haworth, R. D., and Lawton, D. A., *J. chem. Soc.*, 2173 (1963).
61. Grüttner, F., *Arch. Pharm.*, *Berl.*, **236**, 278 (1898).
62. Schmidt, O. T., *Justus Liebigs Annln Chem.*, **476**, 250 (1929); **515**, 43, 65 (1934).
63. Schmidt, O. T., *Fortschr. Chem. org. NatStoffe*, **14**, 71 (1956).
64. Mayer, W., and Kunz, W., *Naturwissenschaften*, **46**, 206 (1959).
65. Kutani, N., *Chem. pharm. Bull.*, *Tokyo*, **8**, 72 (1960).
66. Hathway, D. E., *Biochem. J.*, **63**, 380 (1956).
67. Hillis, W. E., and Carle, A., *Biochem. J.*, **74**, 607 (1960).
68. Schmidt, O. T., and Hersk, J., *Justus Liebigs Annln Chem.*, **587**, 63 (1954).
69. Schmidt, O. T., and Reuss, H., *Justus Liebigs Annln Chem.*, **649**, 137 (1961).
70. Schmidt, O. T., and Schmadel, H., *Justus Liebigs Annln Chem.*, **649**, 157 (1961).
71. Schmidt. O. T., and Schmadel, H., *Justus Liebigs Annln Chem.*, **649**, 149 (1961).
72. Schmidt, O. T., and Schach, A., *Justus Liebigs Annln Chem.*, **571**, 29 (1951).
73. Freudenberg, K., *Ber. dt. chem. Ges.*, **52**, 1238 (1919); **53**, 1728 (1920); *Justus Liebigs Annln Chem.*, **452**, 303 (1927).
74. Schmidt, O. T., Demmler, K., Bittermann, H., and Stephan, P., *Justus Liebigs Annln Chem.*, **609**, 192 (1957).
75. Schmidt, O. T., and Klinger, G., *Justus Liebigs Annln Chem.*, **609**, 199 (1957).
76. Bradfield, A. E., and Penney, M., *J. chem. Soc.*, 2249 (1948).
77. Roberts, E. A. H., and Myers, M., *J. Sci. Fd Agric.*, **3**, 153 (1960).
78. Freudenberg, K., Rein, H. G., and Porter, J., *Justus Liebigs Annln Chem.*, **603**, 177 (1957).
79. Fischer, E., Bergmann, M., and Lipschitz, W., *Ber. dt. chem. Ges.*, **51**, 45 (1918).

80. Crabtree, P., Haslam, E., Haworth, R. D., and Mills, S. D., unpublished observations.
81. Harborne, J., and Corner, J. J., *Biochem. J.*, **81**, 242 (1961).
82. Maier, V. P., and Metzler, D. M., American Chemical Society, 144th meeting, Los Angeles (1963).
83. Scarpati, M. L., and Oriente, G., *Tetrahedron*, **4**, 43 (1958).
84. Harborne, J., *Fortschr. Chem. org. NatStoffe*, **20**, 165 (1962).
85. Herrmann, L., *Pharmazie*, **11**, 433 (1956).
86. Panizzi, L., and Scarpati, M. L., *Gazzetta*, **84**, 792, 806 (1954).
87. Scarpati, M. L., and Esposito, P., *Tetrahedron Lett.*, 1147 (1963).
88. Barnes, H. M., Feldman, J. R., and White, W. V., *J. Am. chem. Soc.*, **72**, 4178 (1950).
89. Uritani, I., and Miyano, M., *Nature, Lond.*, **175**, 812 (1955).
90. Bate-Smith, E. C., *Chemy Ind.* (1956).
91. Erdtman, H., *Svensk. Kem. Tidskr.*, **47**, 223 (1935).
92. Reichel, L., Haussler, R., Pastuska, G., and Schulz, M., *Naturwissenschaften*, **44**, 89 (1957).
93. Herzig, J., Pollak, J., and von Bronneck, M., *Monatsh.* **29**, 278 (1908).
94. Shinoda, J., and Kun, C. P., *J. pharm. Soc. Japan*, **51**, 502 (1930).
95. Briggs, L. H., Cambie, K. C., Lowry, J. B., and Seeley, R. N., *J. chem. Soc.*, 642 (1961).
96. Schmidt, O. T., Komarek, E., and Rentel, H., *Justus Liebigs Annln Chem.*, **602**, 50 (1957).
97. Jurd, L., *J. Am. chem. Soc.*, **81**, 4606 (1959).
98. Schmidt, O. T., and Demmler, K., *Justus Liebigs Annln Chem.*, **576**, 85 (1952); **586**, 179 (1954).
99. Schmidt, O. T., Voigt, H., Puff, W., and Köster, R., *Justus Liebigs Annln Chem.*, **586**, 165 (1954).
100. Schmidt, O. T., and Demmler, K., *Justus Liebigs Annln Chem.*, **587**, 75 (1954).
101. Schmidt, O. T., Blinn, F., and Ladermann, R., *Justus Liebigs Annln Chem.*, **576**, 75 (1952).
102. Schmidt, O. T., "Recent Developments in the Chemistry of Natural Phenolic Compounds", (W. D. Ollis, ed.), p. 146, Pergamon Press, Oxford (1961).
103. Schmidt, O. T., and Grunewald, H. H., *Justus Liebigs Annln Chem.*, **603**, 183 (1957).
104. Schmidt, O. T., and Mayer, W., *Angew. Chem.*, **68**, 103 (1956).
105. Schmidt, O. T., Schmidt, D. M., and Herok, J., *Justus Liebigs Annln Chem.*, **587**, 67 (1954).
106. Schmidt, O. T., and Mayer, W., *Justus Liebigs Annln Chem.*, **571**, 1 (1951).
107. Schmidt, O. T., and Komarek, E., *Justus Liebigs Annln Chem.*, **591**, 156 (1955).
108. Mayer, W., *Justus Liebigs Annln Chem.*, **578**, 34 (1952).
109. Schmidt, O. T., Lademann, R., and Himmele, W., *Chem. Ber.*, **85**, 408 (1952).
110. Schmidt, O. T., Heintzeler, M., and Mayer, W., *Chem. Ber.*, **80**, 510 (1947).
111. Haworth, R. D., Pindred, H., and Jefferies, P., *J. chem. Soc.*, 3617 (1954).
112. Fridolin, A., Ph.D. Thesis, Dorpat, 1884.

113. Schmidt, O. T., Berg, S., and Baer, H. H., *Justus Liebigs Annln Chem.*, **571**, 19 (1951).
114. Schmidt, O. T., and Nieswandt, W., *Justus Liebigs Annln Chem.*, **568**, 165 (1950).
115. Schmidt, O. T., and Schmidt, D. M., *Justus Liebigs Annln Chem.*, **578**, 25 (1952).
116. Schmidt, O. T., Hensler, R. H., and Stephan, P., *Justus Liebigs Annln Chem.*, **609**, 186 (1957).
117. Schmidt, O. T., "Symposium on Vegetable Tannins", Madras (1961); "Meeting of Chemical Society", Sheffield (1962).
118. Schmidt, O. T., and Bernauer, K., *Justus Liebigs Annln Chem.*, **591**, 153 (1955).
119. Haworth, R. D., and Grimshaw, J., *J. chem. Soc.*, **418**, 4225 (1956).
120. Mayer, W., Bachmann, R., and Kraus, F., *Chem. Ber.*, **88**, 316 (1955).
121. Schmidt, O. T., and Eckert, R., *Justus Liebigs Annln Chem.*, **618**, 71 (1958).
122. Mayer, W., *Justus Liebigs Annln Chem.*, **578**, 34 (1952).
123. Schmidt, O. T., and Komarek, E., *Justus Liebigs Annln Chem.*, **591**, 156 (1955).
124. Mayer, W., Fikentscher, R., Schmidt, O. T., and Schmidt, J., *Chem. Ber.*, **93**, 2761 (1960).
125. Mayer, W., and Fikentscher, R., *Chem. Ber.*, **91**, 1542 (1958).
126. Jurd, L., *J. Am. chem. Soc.*, **80**, 2249 (1958).

CHAPTER 5

The Biosynthesis of Plant Polyphenols

I. Introduction

The astonishingly rapid progress made in recent years towards an understanding of the pathways pursued in Nature for the synthesis of many of its secondary products has been due primarily to the application of new experimental techniques such as the use of isotopic tracers and improved enzymic methods. It is perhaps significant that the outcome of these developments has been in many cases to give clear support for the outlines, if not the details, of many of the earlier biogenetic hypotheses[1, 2] which were based usually on comparisons of molecular structure within a group of natural products. This encourages the belief that speculative ideas of this type may still play a useful although perhaps more limited role in the elucidation of pathways of biogenesis.

In 1907 Collie[3] suggested that "acetate" might function as a precursor of certain phenols in Nature and in 1935 H. O. L. Fischer[4] predicted that shikimic (9) and quinic (10) acids might act as intermediates in the formation of the widely distributed plant phenol gallic acid. Various groups of workers have in recent years established experimentally the validity of these ideas and two major pathways—based on acetate[5] and shikimic acid[6] respectively—are now recognized for the synthesis of natural phenolic compounds. The experimental proof for these two pathways has been derived predominantly from work with micro-organisms, in the case of the acetate route with moulds and fungi and for the shikimic acid pathway with radiation induced bacterial mutants. Evidence for the existence of these same synthetic mechanisms in higher plants relies largely on the theory of a unity of metabolic pathways in Nature and on the results of isotopic tracer experiments which are most reasonably explained in the majority of cases by the assumption that the pathways in question operate here also. In the case of synthesis via shikimic acid additional support has been obtained from the discovery of the widespread occurrence in plant tissues of both shikimic (9) and quinic (10) acids and their derivatives[7] and of two important enzymes[8]—5-dehydroquinase and 5-dehydro-shikimic reductase (Fig. 2(a) and (b) respectively)—in this metabolic sequence.

Although the isotopic tracer method has contributed most to our present knowledge of the biosynthesis of plant polyphenols and will undoubtedly lead to further advances in this field of investigation it has its own pitfalls and associated ambiguities which necessitate care in the design and interpretation of experiments. A justifiable criticism may be levelled in certain instances at the assumption that successful incorporation of a substrate into a plant product indicates that it is a true biosynthetic intermediate and not merely a possible precursor. The validity of the approach to the biosynthesis of plant products via the isotopic tracer method is most readily demonstrated when the sequence in a particular pathway predicted from tracer studies is shown in fact to be correct by subsequent work at the enzymic level. An excellent example of this is in the work of Brown[9] and Neish[10] and their collaborators on the biosynthesis of lignin. These workers concluded from tracer studies that the most probable pathways to the substituted cinnamyl alcohols, the accepted precursors of lignin[11-13], are as shown in Fig. 5. An interesting feature of their results was the observation that tyrosine (16) is unable to function as a precursor of lignin except in members of the Gramineae family (e.g. wheat—*Triticum vulgare*), whilst in the same capacity phenylalanine is uniformly efficient in most plant families. The later work of Neish[14] and Conn[15] has shown that these effects are due to the relative distribution in plants of the two deaminases whose function is to convert the aromatic amino acids to the corresponding cinnamic acids. The enzyme acting on phenylalanine was shown to be ubiquitous and present in both mono- and dicotelydons but the tyrosine deaminase (tyrase) was generally but not exclusively limited to members of the Gramineae, thus supporting the biosynthetic evidence deduced earlier from tracer work.

The principal difficulties in the study of biosynthetic pathways in higher plants using isotopic tracers result from the need to work with the whole or a particular part of the plant and hence with enzyme systems which are contained within the cellular structure of the organism. In these circumstances the efficiency of a particular substrate as a biosynthetic precursor to a secondary plant product is dependent not only on its proximity in the metabolic pathway to the compound in question but also on its rate of uptake and transport to the site of enzymic synthesis. In many isotopic tracer experiments using plants these last two factors remain largely unknown quantities and thus comparative measurements of efficiency of incorporation (such as dilution values) may frequently have less than their intended significance. These technical difficulties inherent in the plant system are revealed in many cases by a low percentage incorporation of radioactively labelled substrates. A further interesting illustration of the effects which arise due to difficulties in penetration of substrates through cellular membranes to the site of enzymic synthesis comes from the work of Goodwin[16] in a study of sterol and terpene biosynthesis in the developing chloroplasts of maize seedlings. Goodwin has postulated for this system two groups of enzymes working independently and separated by the semipermeable chloroplast membrane, one within the chloroplast controlling carotenoid and other terpene synthesis, the other outside guiding the formation of sterols, pentacyclic triterpenes and plastoquinones. An unusual manifestation of the operation of this dual enzyme system observed by Goodwin was that mevalonic acid—a well authenticated intermediate in all general terpene biosynthesis—was readily incorporated into sterol fractions (extra chloroplastic) but, presumably due to its inability to penetrate the chloroplast membrane, not into the carotenoid and other terpenoid compounds. Analogous studies at the enzymic level of plant polyphenol biosynthesis have been few. In several plants a controlling factor in the incorporation of substrates into phenolic metabolites has been found to be the maturity of the plant tissue. Thus Towers and his colleagues[17, 18] found that radioactively labelled precursors of the $C_6 \cdot C_3$ type were readily incorporated into arbutin and phloridzin in *Pyrus* and *Malus* leaf discs prepared from young but not old leaves. Similar observations have been made[19] on the incorporation of isotopically labelled glucose and other substrates into Sumach gallotannin in the leaves of *Rhus typhina*. Although these differences may be a reflection of the rate of synthesis of these particular phenols in young and old tissues it may also be argued that in older leaves the substrate has greater difficulty in reaching the required site of synthesis.

The question of aberrant synthesis following the administration of different compounds to a plant is one which is frequently mentioned but one whose importance under the present circumstances is extraordinarily difficult to evaluate. Several observations indicate that plants are able to transform substrates which are not normal intermediates in a biosynthetic sequence into normal and abnormal end products. As an instance of the former case Grisebach and Kellner[20] fed 2-^{14}C-p-fluorocinnamic acid to red cabbage seedlings and found that radioactivity was incorporated into the red pigment cyanidin (35) and to explain this observation it was suggested that the fluoro acid was either first converted to p-coumaric acid or to smaller fragments before incorporation. Alternatively feeding of the various hydroxycinnamic acids (21, 22, 23, 24), accepted precursors of lignin, to plants leads in many cases to the formation of their glucosides or glucose esters[21] and occasionally to $C_6 \cdot C_1$ fragments none of which for a given plant may be identified as normal end products of metabolism.

In general, however, despite these difficulties and uncertainties, the isotopic tracer technique has proved to be a powerful one when applied to the elucidation of pathways of biosynthesis. In the field of plant polyphenols the assumption has been made that the well authenticated acetate and shikimic acid pathways function in plants and in the main workers have been reluctant to seek explanation of their results except within the framework of these accepted pathways of biosynthesis.

II. The Acetate Pathway

Nearly seventy years ago Collie[3] studying the cyclization and condensation reactions of poly-β-keto systems observed the close resemblance of several of the products of naturally occurring phenols and he postulated that similar processes might take place in Nature. His work remained largely forgotten until Birch and Donovan[22] reintroduced the concept in an attempt to extrapolate the use of acetic acid (as acetyl coenzyme A) as a building unit in the biosynthesis of phenols. The hypothesis of Birch and Donovan is based on the multiple condensation of acetate (or malonate)[5] units to give a poly-β-keto system which is then conceived as cyclizing by aldol or Claisen type condensations to produce phenolic compounds having the resorcinol or phloroglucinol orientation of hydroxyl groups (Fig. 1). Carbon–oxygen skeletons generated in this way may, it was postulated, undergo further secondary structural modifications either before or after cyclization and this has enabled the extension of the acetate hypothesis to many natural phenols not, at first sight, directly within its scope. Many examples of such processes have been discussed by workers in this field[5] and the use of labelled acetate as precursor has provided experimental proof of the validity of these ideas. An outstanding problem in acetate biosynthesis has been the inability to define the exact sequence of events leading from acetate (or most probably malonate)[5] to the observed metabolic product. An attractive hypothesis put forward by Birch[5], which has led to fruitful results in the interpretation of intermediate steps in the biosynthesis of several mould metabolites, is that the primer

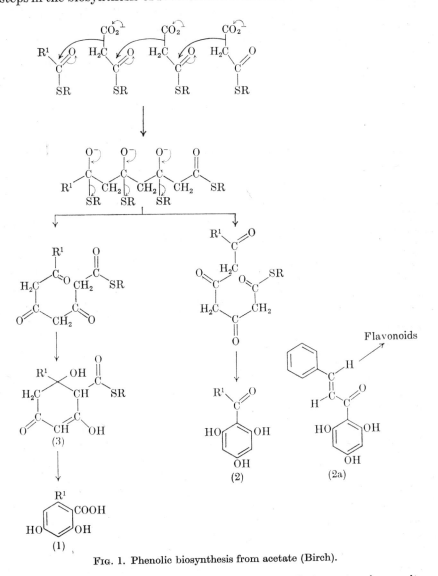

Fig. 1. Phenolic biosynthesis from acetate (Birch).

acid (Fig. 1, R^1—normally acetate) and the chain extension units (malonyl coenzyme A) are attached to the enzyme surface, the latter via their carboxyl groups, and condense together in a concerted manner

(Fig. 1) to yield the poly-β-keto system (the action may be envisaged as analogous to a zip moving along a zip fastener). Steps in the subsequent conversion of the poly-β-keto system to the final phenolic products (1, 2) may produce cyclic non-aromatic intermediates (e.g. 3) which have a separate existence.

Experimental support for the operation of this route of biosynthesis in higher plants has been obtained from studies of the biosynthesis of various flavonoid compounds using isotopically labelled acetate precursors. Experiments by several groups of workers[23-28] have supported the original suggestion of Birch[22] that flavonoid biosynthesis probably proceeds by condensation of a cinnamic acid (Fig. 1. $R^1 \cdot COOH = C_6H_5 \cdot CH = CH \cdot COOH$) or related $C_6 \cdot C_3$ component with acetate or malonate units to give as the first identifiable intermediate the chalcone (2a) which is then converted to the various flavonoids by cyclization and modification of the oxidation state of the heterocyclic ring. These ideas are of obvious importance in any study of the formation of the condensed tannins at least in so far as their probable precursors the flavan-3,4-diols and flavan-3-ols are concerned. Prior to a fuller discussion of the problems of flavonoid biosynthesis it is necessary to outline the shikimic acid pathway of aromatic biosynthesis, which leads not only to the $C_6 \cdot C_3$ units such as the cinnamic acids involved in lignin and flavonoid biosynthesis but also to gallic acid a principal component of the gallotannins and in a modified form of the ellagitannins.

III. THE SHIKIMIC ACID PATHWAY

The elegant studies of Davies,[29] Sprinson,[6] Gibson[30, 31] and their collaborators using mutants of *Escherichia coli* and *Aerobacter aerogenes* have revealed an important pathway of aromatic biosynthesis leading to phenolic compounds with a $C_6 \cdot C_3$ or $C_6 \cdot C_1$ carbon skeleton and with predominantly an ortho disposition of the hydroxyl groups in the aromatic ring system. The route, which is referred to as the shikimic or prephenic acid pathway, has been adequately reviewed by Sprinson[6] and the various steps originating with the carbohydrate source are shown in Fig. 2. Recent investigations have concerned details of certain steps in the pathway which have hitherto remained obscure, in particular those involving the cyclization of 3-deoxy-D-*arabino*-heptulosonic acid-7-phosphate (6) to 5-dehydroquinic acid (7) and the conversion of shikimic acid-5-phosphate (11) to prephenic acid (14).

Sprinson[6] and his colleagues have made several observations in connection with the former reaction and have shown it to be Co^{2+} and

NAD dependent. Attempts, however, to accumulate intermediates by omission of one or both of these co-factors were unsuccessful and Sprinson[6] has put forward the mechanism in Fig. 3 for this change. By comparison considerable clarification of the steps immediately prior to the formation of prephenic acid (14) has been possible. The three carbon

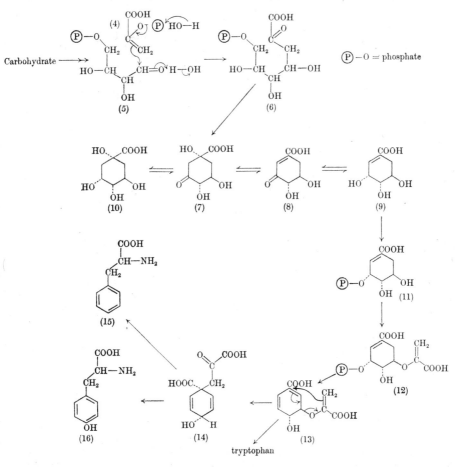

FIG. 2. The shikimic acid pathway.

side chain of pyruvic acid is envisaged as being attached to C_1 of shikimic acid by a unique series of reactions in which shikimic acid 5-phosphate (11) is converted via the enol pyruvate derivative (12) and thence by elimination of phosphate to chorismic acid (13) isolated by Gibson and Gibson[30] from a strain of *Acrobacter aerogenes*. Chorismic

6+

acid represents a branch point in the pathway leading on the one hand to phenylalanine (15) and tyrosine (16) and on the other via anthranilic acid to tryptophan. The structure (13) assigned to chorismic acid[31] is based on its spectroscopic properties, on its conversion by warming in dilute alkaline solution to prephenic, phenylpyruvic and *p*-hydroxy-benzoic acids and on its enzymic transformation to anthranilic acid

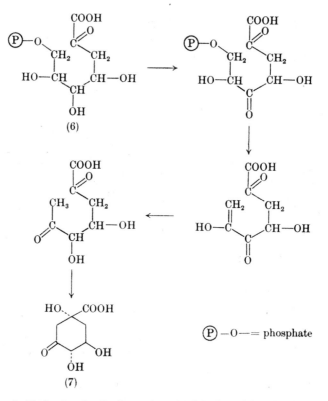

Fig. 3. Mechanism for the formation of 5-dehydroquinic acid (Sprinson).

when one mole of pyruvic acid is liberated. The formation of prephenic acid (14) from chorismic acid (13) has been formulated as an SNi displacement reaction and leads to a structure with a *cis* relationship between the carboxyl group at position 1 and the hydroxyl group at position 4. Prephenic acid has a half life of 60 sec in N acid solution[32] but is relatively more stable at neutral pH and has been isolated and characterized as its barium salt. An unequivocal synthesis of this remarkably labile intermediate has not been reported but Plieninger

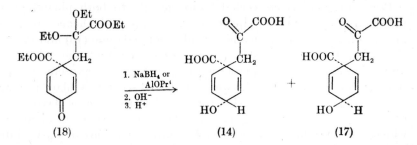

(18)　　　　　　　　(14)　　　　　　　　(17)

and his collaborators have confirmed its stereochemistry in a synthesis
of tetrahydroprephenic acid [33] and prepared a mixture of prephenic (14)
and epiprephenic (17) acids by the action of alkali followed by acid on
the reduction products of the dienone (18)[34, 35].

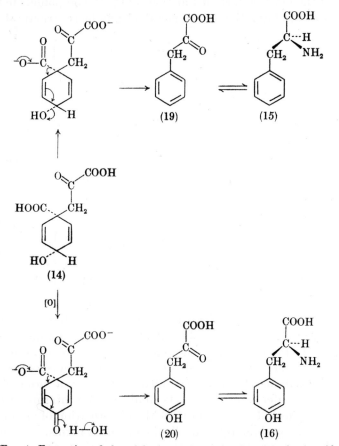

FIG. 4. Formation of phenylalanine and tyrosine from prephenic acid.

Prephenic acid is converted[36] enzymically to both phenylpyruvic (19) and *p*-hydroxyphenylpyruvic (20) acids and thence by transamination to the aromatic amino acids phenylalanine (15) and tyrosine (16) (Fig. 4). Various workers have suggested that the role of the aromatic amino acids in plants is to act not only in protein synthesis but as a reserve source of the wide number of secondary plant products possessing a $C_6 \cdot C_3$ carbon skeleton[37] and those such as flavonoids[22], coumarins[10], stilbenes[22], lignin[9] and the several classes of alkaloids[38] which are thought to be derived by transformations of $C_6 \cdot C_3$ intermediates.

IV. METABOLISM OF $C_6 \cdot C_3$ COMPOUNDS IN PLANTS

One of the most important functions of $C_6 \cdot C_3$ compounds in vascular plants is in the formation of the metabolically inert structural polymer

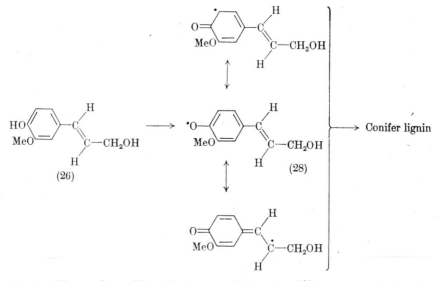

lignin. The studies of Freudenberg and his school[39] have revealed quite clearly the pattern of the polymerization processes which constitute lignification. Freudenberg[40] showed that a phenol oxidase, *laccase*, derived from the expressed juices of the common mushroom *Psalliota campestris* polymerizes coniferyl alcohol (26) to a product closely resembling conifer lignin and acts similarly on a mixture of coniferyl and sinapyl alcohols (26) and (27) with the formation of an amorphous copolymer very similar to the lignin of angiosperms. A free radical mechanism was proposed[40] for the polymerization process in lignification in which the central intermediate in the formation of coniferyl polymers is the mesomeric radical (28) resulting from the oxidation of the coniferyl alcohol precursor (26).

Neish, Brown[11-13] and Nord[41] and their respective collaborators have made extensive isotopic tracer studies on the biosynthesis of lignin with particular reference to the identification of pathways from a carbohydrate source to the cinnamyl alcohols (25, 26 and 27) the recognized precursors of lignin. Brown[9] and Neish[10] concluded as a result of their studies that following prephenic acid the most probable pathways to the substituted cinnamyl alcohols are as shown in Fig. 5. Essentially the methods used by Neish and Brown utilized the availability of isotopically labelled $C_6 \cdot C_3$ precursors of various types and compared the efficiency with which they acted as precursors of lignin. Subsequent work at the enzymic level by Neish[14] and Conn[15] has substantiated various features of these pathways in particular the differing roles of the aromatic amino acids phenylalanine (15) and tyrosine (16).

The metabolic sequences from prephenic acid to lignin outlined by Neish and Brown represent the most rational explanation of their numerous feeding experiments. The work is, however, notable for the fact that several compounds were incorporated into lignin in a manner not readily explicable in terms of the above pathways (Fig. 5) and in these cases minor pathways of lignin biosynthesis were postulated as occurring. Thus although $C_6 \cdot C_1$ compounds were in general poor precursors of lignin when fed to wheat plants[11] vanillin was incorporated into wheat lignin with an efficiency varying from 20–30% of that of phenylalanine and in this case it was suggested that the initial step in the biosynthesis involved a union of a C_2 fragment with vanillin to give a precursor of the $C_6 \cdot C_3$ type[9]. Similarly Brown and Higuchi[13] have shown the ready assimilation of sinapic acid (24) into both the *coniferyl* (26) and sinapyl (27) components of wheat lignin, the ratio of incorporation varying from 1 to 6 in young seedlings (25 days) to 1 to 16 in heading wheat (73 days). To account for these unusual results Brown and Higuchi[13] postulated that the young wheat contained a powerful demethoxylation system capable presumably of converting the sinapyl to the coniferyl grouping. In both of these cases, however, it would also seem reasonable to argue that although the wheat plant is able to utilize both substrates in lignin formation under the experimental conditions of feeding, under the usual circumstances of growth these are not the normally adopted metabolic paths.

The observations in the work of Neish[10] and Brown[9] on lignin biosynthesis that phenylalanine and tyrosine may well have non-equivalent functions in the synthesis of secondary plant products has been borne

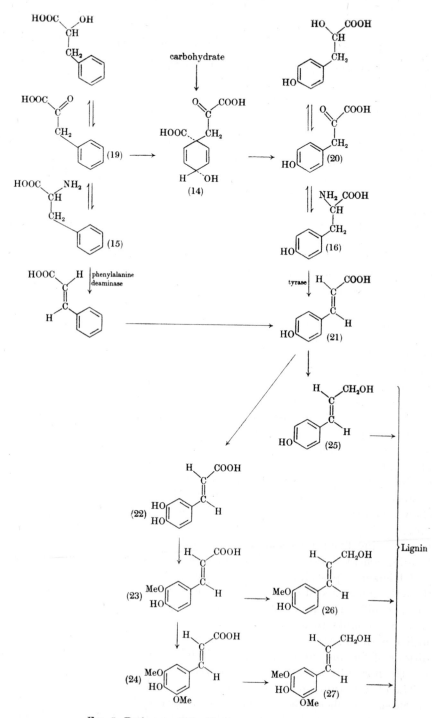

FIG. 5. Pathways of lignification (Neish and Brown).

out in related biosynthetic investigations. In the studies of the biosynthesis of the Amaryllidaceae alkaloids[42, 43] (e.g. haemanthamine
(29), *Haemanthus natalensis*) isotopic tracer experiments showed ring A
to be derived from phenylalanine and ring C and the C_2 chain from
tyrosine. Similarly with colchicine[44] (30) (*Colchicum autumnale*)
analogous experiments demonstrated that ring A and carbon atoms 5,
6 and 7 are formed from the phenylalanine–cinnamic acid pathway and
suggested that the tropolone ring (C) is formed by ring expansion of a
C_6–C_1 unit derived from tyrosine.

A noteworthy structural feature amongst naturally occurring phenylpropanoid compounds and substances derived biogenetically from them

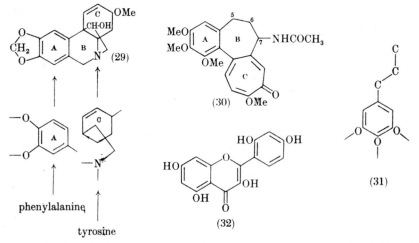

is the hydroxylation pattern in the aromatic ring system. Hydroxyl,
methoxyl and methylenedioxy groups are found predominantly, although not exclusively, in the 3, 4 and 5 positions relative to the C_3
side chain (31). This same pattern of oxygenation, with the exception
of a few isolated examples such as the pigments of *Artocarpus integrifolia* heartwood (e.g. morin (32)) is common also to ring B of the
flavonoids and is assumed to result, as with the $C_6 \cdot C_3$ compounds
themselves, from a direct and specific hydroxylation of the aromatic
ring system. Enzyme systems have been isolated from plants which
have this specific property of hydroxylation of the aromatic ring
system[45] and isotopic tracer experiments suggest that the formation
of these particular oxygenated phenylpropanoid compounds and their
derivatives is associated in many plants with the metabolism of phenylalanine and tyrosine (for example the biosynthesis of lignin in dico-
tyledons and of colchicine above). In the structure of many of the

isoflavonoids and the rotenoids an additional structural feature which may have biogenetic significance is the oxygen atom in the 2′ position in ring B. Coumarin derivatives similarly possess an oxygen function ortho to the C_3 chain and in 1942 Haworth[46] suggested that the introduction of this group could occur by direct oxidative cyclization of a cinnamic acid derivative. Later workers have developed the idea[47] and Kenner and his group[48], in a study of the biosynthesis of the mould metabolite novobiocin, obtained valuable evidence in support of this mechanism. However, feeding experiments to determine the pathway of biosynthesis of coumarin (33; R = H) in *Melilotus alba*[49], of herniarin (33; R = OMe) in *Lavandula officinalis*[50] and umbelliferone (33;

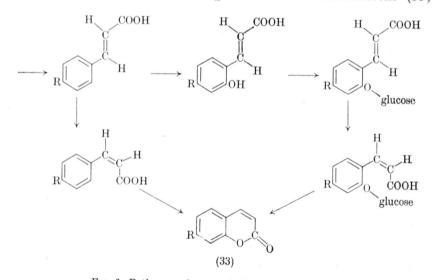

FIG. 6. Pathways of coumarin biosynthesis [46, 49–51].

R = OH) in *Hydrangea macrophylla*[51] accord most satisfactorily with a biosynthetic sequence in which an enzymically controlled ortho-hydroxylation of the *trans*-cinnamic acid occurs (Fig. 6).

   The experimental evidence available demonstrates quite clearly therefore that the acetate and shikimic acid pathways represent two contrasting routes to natural phenols. In the former pathway the majority of the phenolic hydroxyl groups in the product are generated directly from the original oxygen atoms in the acetate or malonate building units whereas direct nuclear hydroxylation appears to be the predominant manner in which phenols of the $C_6 \cdot C_3$ type are formed. This latter method is one perhaps which the organic chemist would not have predicted from an analysis of the structural anatomy of these compounds and consideration of the fact that they are derived from shikimic acid which already possesses the "desired" oxygenation pattern.

V. BIOSYNTHESIS OF FLAVONOIDS AND CONDENSED TANNINS

The metabolism of phenyl propanoid compounds in vascular plants is intimately connected not only with lignin biosynthesis but also with that of a range of other substances notably the flavonoid group of natural products. The various classes of flavonoids and compounds such as the rotenoids and isoflavonoids which bear a close structural relationship to the flavonoids may be considered to be derived basically from a

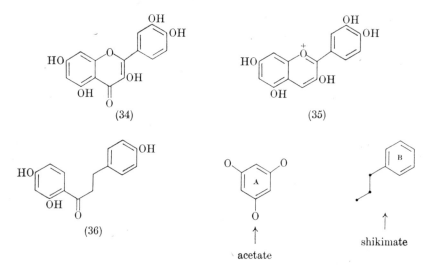

(34)

(35)

(36)

acetate

shikimate

1,3-diarylpropane (C_6—C_3—C_6) unit which Robinson[52] first suggested could be formed in nature by condensation of a C_6 (phloroglucinol) and a $C_6 \cdot C_3$ (catechol) fragment. The subsequent ideas of Birch and his collaborators[22] in the field of phenolic biosynthesis from acetate made possible a plausible rationale along these lines. Birch[5, 22] suggested that condensation of a cinnamic acid (or related unit) and three malonate molecules, as envisaged in Fig. 1 (R^1 = —CH=CH—⟨⟩), would yield the chalcone (2a) as the most elementary compound of the C_6—C_3—C_6 type in flavonoid biosynthesis. Isotopic tracer studies have demonstrated that the production of quercetin (34) in buckwheat (*Fagopyrum tataricum* and *F. esculentum*)[23, 24, 26], of cyanidin (35) in

6*

red cabbage and buckwheat (*F. esculentum*)[26–28], of phloridzin (36) in apple leaf (*Malus* sp.)[18] and of other flavonoids conform to this general hypothesis in which the C_6 unit (Ring A) is acetate derived and that the C_6—C_3 unit (ring B) has its origin in the shikimic acid pathway.

The classification of flavonoids into major groups (flavonols, 2,3-dihydroflavonols, flavan-3-ols, flavan-3,4-diols, anthocyanidins, etc.), is

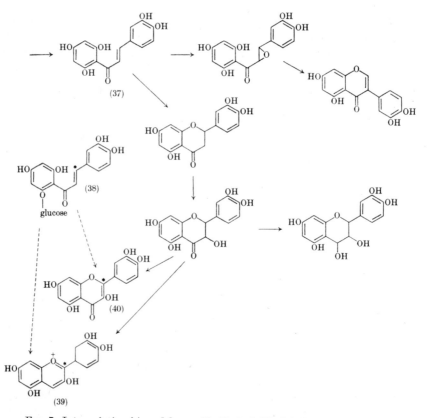

FIG. 7. Inter-relationships of flavonoids (Seshadri[53], Grisebach and Ollis[47]).

based on the configuration and state of oxidation of the C_3 unit in the molecule. In the majority of flavonoids this unit is involved in the formation of a 6- (or in the case of aurones a 5-) membered oxygen heterocyclic ring, although it remains uncyclized in the chalcones and their dihydroderivatives. The biosynthesis of the individual types of flavonoid which is of interest in regard to the derivation of the flavan-3-ols and flavan-3,4-diols the suggested precursors of the condensed tannins has been a subject of some speculation. Many workers consider that they are sequentially synthesized from a single C_{15} precursor such as the chalcone (2a). Such theories are based generally on the co-occurrence in a single plant species of different flavonoid types with similar patterns of oxygenation in the aromatic nuclei (A and B). Schemes of an entirely different character have resulted from the use of such arguments, thus Seshadri[53] favoured a scheme, which has also been supported in principle by Grisebach and Ollis[47], in which the chalcone (37) is the first formed C_6—C_3—C_6 compound in flavonoid biogenesis (Fig. 7). Grisebach and Patschke[25] have produced experimental evidence in support of this hypothesis, thus the isotopically labelled chalcone (38) when fed to red cabbage seedlings and buckwheat respectively gave cyanidin (39) and quercetin (40) respectively with specific incorporation of the labelled atom at the position (∗) shown. These experiments did not clearly rule out the possibility of breakdown of the chalcone (38) into simpler fragments prior to incorporation into the flavonoids and final proof of the validity of this pathway must await the results of feeding experiments with doubly labelled chalcone precursors and experiments at an enzymic level.

The biosynthesis of the isoflavones, and the 3- and 4-phenylcoumarins present interesting biogenetic problems since although they are obviously related to the simple flavonoids, ring B and carbon atoms 2, 3 and 4 in the heterocyclic ring cannot obviously be derived from an intact $C_6 \cdot C_3$ precursor. Grisebach and his colleagues[25] have carried out intensive tracer studies on the biosynthesis of isoflavones and these point to a 1,2-aryl migration occurring following the formation of the C_6—C_3—C_6 intermediate. Grisebach[54] has formulated the change in terms of a rearrangement of the aryl group in the chalcone epoxide (Fig. 7). The relationship of the isoflavones to the rotenoids has been discussed by several workers and Grisebach and Ollis[47] have proposed possible routes to the rotenoids and the closely related 2′-oxygenated isoflavones from phenyl propanoid precursors hydroxylated in the 2 position. The introduction of isoprenoid residues into flavonoid compounds and phenols in general has been suggested by Birch as involving processes of C-alkylation and the idea has been discussed and exemplified in detail by Ollis and Sutherland[55] in a recent review.

Roux, following a study of the distribution of flavonoid compounds in various heartwoods (*Schinopsis* species[56], *Acacia mollisima*[57] and *Robinia pseudacacia*[58]), has put forward a radically different picture of flavonoid biosynthesis in which the flavan-3,4-diol (41) is considered to hold a central position (Fig. 8). The flavan-3,4-diol he suggested is first formed and subsequently converted to the other flavonoids by a combination of reduction and dehydrogenation reactions.

The alternative proposal that flavonoid biosynthesis proceeds along pathways which vary according to the class of flavonoid has received little attention, although Robinson[52] originally suggested that the flavones and anthocyanidins were produced divergently from two intermediates, one of which was present in limited amount but was common to both groups of pigments—a C_6 fragment constituting ring A—and the second—a $C_6 \cdot C_3$ unit giving ring B and the C_3 linkage—which was produced in amount and variety dependent on environmental factors. A speculative scheme which embodies the concept of diverse pathways

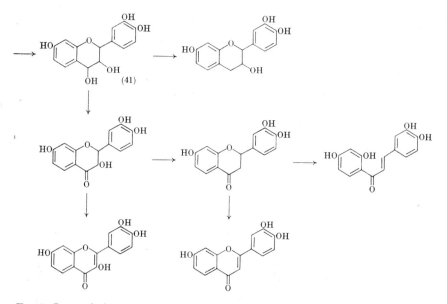

FIG. 8. Inter-relationships of flavonoid constituents of *Acacia mollisima* (Roux)[56–58].

to different flavonoids and which provides a more rational explanation of the origin of the 3-hydroxyl group in flavonols, 2,3-dihydroflavonols, etc., is shown in Fig. 9. Further experimental evidence, probably at the enzymic level, is, however, clearly necessary to elucidate these finer points of flavonoid biosynthesis in plant tissues.

The consensus of opinion favours the proposal that condensed tannins are derived by oxidative or simulated acid catalysed condensation of flavan-3-ols and flavan-3,4-diols and the mechanisms proposed for these conversions have been considered in Chapter 3. Whether in all parts of the plant these reactions represent enzymically controlled biosynthetic or post mortal processes remains open to question. In this context the question of the site of synthesis of heartwood polyphenols

has aroused some interest and many workers consider that polyphenols
are initially formed in the leaves and then translocated to other parts
of the plant in much the same way as carbohydrates and amino acids.
Swain and Hillis[59] in a study of the metabolism of phenolic compounds
in *Prunus domestica* showed that the quantity of leucoanthocyanins (of
unspecified chemical structure) in young leaves increased rapidly as

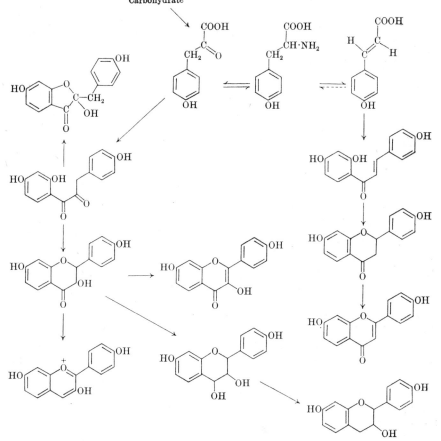

FIG. 9. Further possible pathways of flavonoid biosynthesis.

the leaves matured to maximum size when the phenolic content of the
leaves took up an almost stationary value. This evidence they took to
support the view of a "steady state" in phenol synthesis in which the
phenols once synthesized were metabolized or translocated away from
the leaves to other tissues. Hillis[60], however, in a later review has stated
that translocation of polyphenols or their alicyclic precursors does not

occur to any appreciable extent in woody tissues and that those present in the woody part of a plant are formed within the living cells of the phloem, sapwood and cambium. This problem of location of polyphenol synthesis in woody plants is obviously of some importance in regard to a study—if this is ever possible—of the biogenesis of the heartwood and bark polyphenols, pre-eminent among which are the condensed or non-hydrolysable tannins, and there is obviously still considerable scope for detailed work which is necessary to fill the gap in our knowledge regarding the formation of these important plant products.

VI. BIOSYNTHESIS OF HYDROLYSABLE TANNINS

In contrast to the formation of condensed tannins in plants there is little doubt that the hydrolysable tannins are direct products of plant metabolism and as such they are therefore more amenable to a normal biosynthetic study. Gallic acid (42) and its oxidation product hexahydroxydiphenic acid (43) which are the principal phenolic constituents of the gallotannins and ellagitannins respectively are normally encountered in plant tissues in the form of esters with sugars or related substances and in the case of gallic acid also with the flavan-3-ols. (−)-epicatechin and (−)-epigallocatechin. Many of the numerous reports in the literature claiming the presence of *free* gallic acid in plant tissues may in part have resulted from enzymic or chemical decomposition of the labile gallate esters during extraction and although careful analysis has shown the acid to occur free in some plants it is most usually found in ester form. In lower organisms there is good evidence for the existence of gallic acid in the free state. Albrecht and Bernard[61] isolated the acid along with the related protocatechuic acid from the culture medium of *Phycomyces blakesleeanus* grown on a glucose medium and several studies of the biosynthesis of gallic acid have concentrated on its formation by this particular organism. Brucker[62, 63] showed that inositol was not a normal precursor of gallic or protocatechuic acids but that an increase in the concentration of various amino acids such as tyrosine in the growth medium led to an increased production of both phenolic acids by the mould. When 6-^{14}C-D-glucose was fed to the mould the specific activities of the isolated shikimic, protocatechuic and gallic acids were respectively 80, 67 and 12% of that of the substrate and in a similar experiment isotopically labelled shikimic acid was readily incorporated into protocatechuic but not gallic acid. From these observations Brucker[63] concluded that shikimic acid is an intermediate between glucose and protocatechuic acid but that gallic acid may arise from simpler building units or by aromatic amino acid metabolism. In higher plants oxidative degradation of $C_6 \cdot C_3 \rightarrow C_6 \cdot C_1$ compounds has been shown to be a pathway frequently utilized in the formation of secondary plant products[64-66]. Thus methyl salicylate[64]—a component of the young leaves of *Gaultheria procumbens*—has been shown to be derived by β-oxidation of the C_3 side chain and hydroxylation of the aromatic nucleus of cinnamic acid. Similarly several groups of workers have demonstrated that phenylalanine and cinnamic, *p*-coumaric and caffeic acids may all act as precursors of the $C_6 \cdot C_1$ unit which is built into the

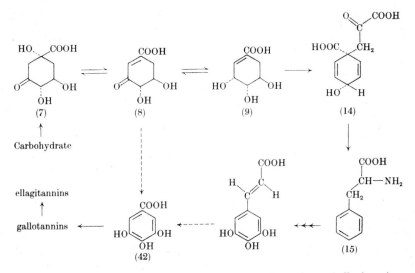

FIG. 10. Possible pathways of biosynthesis of the gallotannins and ellagitannins.

Amaryllidaceae alkaloids such as haemanthamine (27). Zenk[67] in a study of the biosynthesis of gallic acid in *Rhus typhina* concluded from isotpic feeding experiments that this essential component of the gallotannin structure is derived by metabolism of phenylalanine as shown in Fig. 10, and in a manner which therefore conforms to this general pattern of breakdown of $C_6 \cdot C_3$ to $C_6 \cdot C_1$ compounds in plant tissues.

An alternative suggestion concerning the origin of gallic acid in Nature comes from the work of Conn and Swain[68] in *Geranium pyrenaicum* and of Haworth and his collaborators[69] with the mould *Phycomyces blakesleeanus*. Conn and Swain[68], from a limited number of isotopic feeding experiments, reasoned that gallic acid was more likely to be formed directly from a compound such as shikimic acid (9) than by β-oxidation of a $C_6 \cdot C_3$ compound. Haslam, Haworth and Knowles[69] showed that growth of the mould *Phycomyces blakesleeanus* on a glucose medium supplemented with 5-dehydroquinic (7), 5-dehydroshikimic (8) or shikimic (9) acid resulted in increased gallic acid production; that with 5-dehydroshikimic acid (8) giving the greatest increment. Medium replacement experiments provided further evidence to support the suggestion that 5-dehydroshikimic acid (8) was the immediate and 5-dehydroquinic (7) and shikimic (9) acids near precursors of both gallic and protocatechuic acid in the mould. Confirmation of this proposal came from the ready *in vitro* oxidation of (8) to gallic acid. Gross[70] showed by a study of phenolic metabolism in a mutant strain of *Neurospora crassa* that protocatechuic acid is formed in this organism by dehydration of 5-dehydroshikimic acid (8). The organism required phenylalanine, tyrosine, tryptophan and *p*-aminobenzoic acid in combination or shikimic acid (9) alone for growth. However, the enzymic conversion of (8) to (9) was blocked in the organism's metabolism and as a consequence its culture medium accumulated substantial quantities of both (8) and its dehydration product protocatechuic acid. On the basis of the observations of Haslam, Haworth and Knowles it is possible that a similar situation to the one in the *Neurospora* mutant investigated by Gross[70] appertains in *Phycomyces blakesleeanus* and that the aromatic metabolic pathway is blocked or partially blocked at some stage beyond shikimic acid. As (8) builds up in concentration it is converted by dehydration to protocatechuic acid or dehydrogenation to gallic acid. Although in principle it is possible that more than one biosynthetic route operates for the formation of a natural product further work is desirable to evaluate the relative importance of these two routes to gallic acid (Fig. 10).

The naturally occurring galloyl esters and the gallotannins themselves are then assumed to be derived by esterification of the appropriate alcohol with gallic acid. Plants differ widely in their ability to produce simple or complex galloyl esters, mono esters with D-glucose (β-D-glucogallin) and quinic acid (theogallin) occur frequently but the complex gallotannins containing seven or eight galloyl groups per carbohydrate molecule have a more limited distribution[71], and this appears to be associated with the possession of an enzyme system capable of forming depsidically linked galloyl esters. These variations and those in the nature of the alcohol to which the gallic acid is esterified appear worthy of a more systematic investigation than has yet been attempted.

The ellagitannins are characterized by the deposition of ellagic acid (44) on hydrolysis. Extensive degradative studies on these tannins have shown[72] that ellagic acid is formed following hydrolytic cleavage of (+) or (−)-hexahydroxydiphenic acid from the molecule and

Schmidt[73] considers it improbable, in view of the instability of the free hexahydroxydiphenic acid, that this is preformed in the plant and then esterified to the carbohydrate. It is thought more reasonable that the hexahydroxydiphenoyl group (45) is produced by intramolecular de-

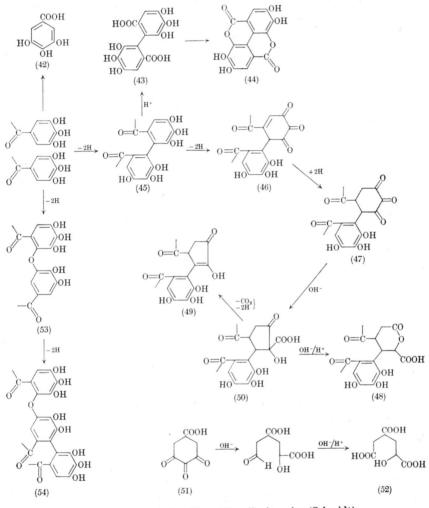

FIG. 11. Inter-relationships of the ellagitannins (Schmidt).

hydrogenation of two galloyl groups in a gallotannin which could be readily brought into the correct steric relationship to one another. Model experiments in the laboratory whose aim was to simulate this reaction have met with varied success. Reichel, Haussler, Pastuska and Schulz[74] demonstrated that photo-oxidation at a neutral pH readily transformed a range of gallate esters to ellagic acid (44). These workers were unable to convert under similar conditions 3,6-di-O-galloyl glucose to the corresponding hexahydroxydiphenoyl derivative and suggested that for this reaction to occur more than 2 galloyl groups per molecule were required. In contrast anodic oxidation of 3,6-di-O-galloyl-1,2-O-isopropylidene-D-glucose and 1,4-di-O-galloyl butane-1,4-diol has been claimed[75] to give the hexahydroxydiphenoyl derivatives in support of the proposed biosynthetic mechanism.

Schmidt and Mayer[73] have extended this hypothesis for the bio-

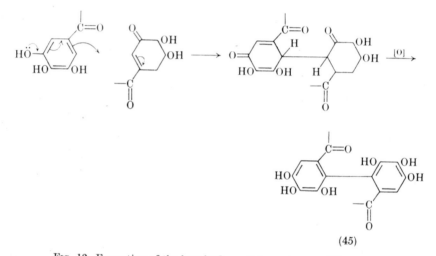

(45)

FIG. 12. Formation of the hexahydroxydiphenoyl group (Wenkert)[78].

genesis of the ellagitannins and have suggested in an elegant scheme (Fig. 11) pathways from the initially formed hexahydroxydiphenoyl group (45) to the numerous complex phenolic acids frequently associated with many of the ellagitannin structures. Formation of the dehydro-hexahydroxydiphenoyl group (46) found in brevilagin I and II is suggested as occurring by oxidation of one of the aromatic ring system, S, subsequent reduction of this ring gives the unusual isohexahydroxy-diphenoyl system (47) which forms part of the structure of terchebin from Myrobalans, and which is thought to be a key intermediate leading both to chebulic acid (48) and brevifolin carboxylic acid (49) in their bound forms. Both would result from a benzylic acid rearrangement of (47) to (50) followed either by hydrolytic cleavage to give (48) or oxidative decarboxylation to give (49). These inter-relationships have received excellent support from the model studies by Mayer, Bachmann and Kraus[76] on the reactions of (51) obtained by reduction followed by oxidation of gallic acid. The acid (51) was readily iso-merized to gallic acid but in cold dilute alkali it rearranged to 2-hydroxy-4-carboxyadipic acid (52) in a manner which is clearly related to the proposed route of formation of chebulic acid (48) from (47). These projected conversions of the alicyclic ring system (47) of terchebin have also been observed paper chromatographically when terchebin was treated with base[77]. All these proposed transformations have been envisaged by Schmidt and Mayer as occurring with the hexahydroxy-diphenoyl group attached to the sugar molecule and it is for this reason that Schmidt has put forward structures for chebulagic and chebulinic acids with the chebulic acid bound to the glucose in a form (48) quite different from that which occurs in the free state.

The elaboration of the dehydrodigalloyl (53) and valoneaic acid (54) residues in ellagitannins have been suggested by Schmidt[73] as occurring by analogous oxidative coupling of galloyl residues but with the formation of carbon–oxygen linkages.

An alternative viewpoint regarding the generation of diphenyl, diphenyl ether and alicyclic $C_6 \cdot C_1$ systems common to many groups of natural products including the ellagitannins has been expressed by Wenkert[78]. These groupings he considered may arise by carbohydrate type condensation and cleavage reactions of hydroaromatic precursors. Thus the hexahydroxydiphenoyl group (45) he suggested was formed by a Michael addition of a galloyl to a 5-dehydroshikimoyl group followed by oxidation (Fig. 12). Further standard oxidation and cleavage reactions would lead to the other characteristic ellagitannin groupings.

VII. ASPECTS OF THE CHEMOTAXONOMY OF THE HYDROLYSABLE TANNINS

Early in the development of natural product chemistry it occurred to many botanists and chemists that it should be possible to bring about a systematic classification of plants on the basis of their chemical con-stituents. Chemotaxonomic studies in principle consist in the investiga-tion of the patterns of compounds occurring in plants (preferably in various parts of the plant—wood, leaves, root, etc.) to obtain evidence for the relationship or non-relationship of plants. The numerous simple and rapid techniques now available for the analysis of plant constituents indicate the probability of considerable progress in this type of study in future years.

Esters and glycosides of a number of hydroxy- and methoxy-benzoic

acids have been observed [65] in all higher plants and preliminary investigations [79] have suggested that gallic acid may have some interest in the
context of chemical taxonomy. Paper chromatographic analysis of the
phenolic constituents of the leaves of fifteen species of the plant family

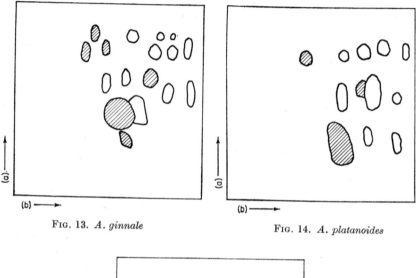

FIG. 13. *A. ginnale*

FIG. 14. *A. platanoides*

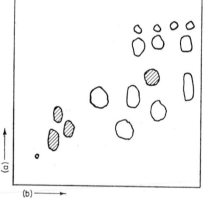

FIG. 15. *A. saccharum*

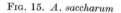

Shaded areas on the chromatograms represent galloyl esters. Solvent systems:
(a) 6% acetic acid; (b) butan-2-ol–acetic acid–water, 14:1:5.

Aceraceae indicated that a subdivision of the plant family was possible
on the basis of the particular form in which gallic acid was "bound"
in the leaves. In 1922 Perkin and Uyeda [80] isolated *Acer* tannin from

the leaves of the Korean maple (*Acer ginnale*) and in later work Kutani[81] assigned to it the structure of 3,6-di-*O*-galloyl-1,5-anhydro-D-glucitol. Paper chromatographic analysis of the phenolic constituents of the leaves of *A. ginnale* gave the pattern shown in Fig. 13. Analysis of the leaves of a further fourteen species of *Acer* showed that another two (*A. tartaricum* and *A. saccharinum*) gave rise to a phenolic pattern very similar to that of *A. ginnale* with 3,6-di-*O*-galloyl-1,5-anhydro-D-glucitol (Fig. 13) as the predominant component containing gallic acid, and along with *A. ginnale* these have been arranged together to form group A. Another three species (group B : *A. platanoides*, *A. campestre* and *A. rubrum*) showed a different distribution of phenolic compounds containing a type of galloyl ester whose ultra-violet absorption and chromatographic properties resembled closely those of Sumach and Chinese gallotannins (Fig. 14, *A. platanoides*). Methyl gallate was rapidly formed when these crude extracts were subject to methanolysis at pH 5·9 thus lending support to the suggestion that in these extracts compounds were present in which gallic acid is bound in the depside form. The major group (group C) of eight species (*A. spicatum*, *A. pennsylvanicum*, *A. rotundilobum*, *A. griseum*, *A. monspessulanum*, *A. saccharum*, *A. palmatum* and *A. pseudoplatanus*) produced in small quantity a series of galloyl esters of unknown structure but distinct paper chromatographic pattern (Fig. 15, *A. saccharum*). The structures of these substances has not been elucidated but their paper chromatographic pattern resembled that of other ellagitannins. Analysis of *A. macrophylla* at various stages of growth showed the leaves of this species to produce very little gallic acid and significantly the formation of hydroxycinnamoyl esters by this species was considerably greater than in the others analysed. Certain areas of agreement between this preliminary subdivision of the *Aceraceae* based on particular phenolic constituents and classifications based on morphological characteristics are apparent. Thus in the classification of Pojarkova[82] *A. campestre* and *A. platanoides* are grouped under *Platanoidea* as are *A. ginnale* and *A. tartaricum* in the sub-group *Trilobata* and both of these relationships also followed from the chemical analysis outlined above. Clearly, however, on this limited sample of *Acer* species (over 150 are recorded)[82] the chemical method is as yet not sufficiently refined to bring out for instance the differences between individual members of group C and show the relationships between *A. rubrum* and *A. saccharinum* to which morphological methods draw attention.

Further work is obviously required in this promising field of investigation to ascertain the true value of gallic acid as a chemotaxonomic tracer. It is nevertheless inevitable that studies along these lines will not only be of value in chemotaxonomy itself but will help to clarify the position in phenolic biosynthesis of this very important metabolite gallic acid.

References

1. Ruzicka, L., *Proc. chem. Soc.*, 341 (1959).
2. Robinson, R., *J. chem. Soc.*, 876 (1917).
3. Collie, J. N., *J. chem. Soc.*, 329 (1893); 1806 (1907).
4. Fischer, H. O. L., and Dangschat, G., *Helv. chem. acta*, **18**, 1206 (1935).
5. Birch, A. J., *Proc. chem. Soc.*, 3 (1962).
6. Sprinson, D. B., *Adv. in Carbohyd. Chem.*, **15**, 235 (1961).
7. Hasegawa, M., "Wood Extractives". (W. E. Hillis, ed.), p. 263, Academic Press, London (1962).
8. Davies, D. D., and Balinsky, D., *Biochem. J.*, **80**, 292, 296, 300 (1961).
9. Brown, S. A., *Science*, **34**, 305 (1961).
10. Neish, A. C., *A. Rev. Pl. Physiol.*, **11**, 55 (1960).
11. Brown, S. A., and Neish, A. C., *Can. J. Biochem. Physiol.*, **33**, 948 (1955).
12. Brown, S. A., and Neish, A. C., *Can. J. Biochem. Physiol.*, **34**, 769 (1956).
13. Brown, S. A., and Higuchi, T., *Can. J. Biochem. Physiol.*, **41**, 613 (1963).
14. Neish, A. C., *Phytochemistry*, **1**, 1 (1962).
15. Koukol, J., and Conn, E. E., *J. Biol. Chem.*, **236**, 2692 (1961).
16. Goodwin, T. W., and Mercer, E. I., "The Control of Lipid Metabolism" (J. K. Grant, ed.), p. 37, Academic Press, London (1963).
17. Grisdale, S. K., and Towers, G. H. N., *Nature, Lond.*, **188**, 1130 (1960).
18. Avadhani, P. N., and Towers, G. H. N., *Can. J. Biochem. Physiol.*, **39**, 1605 (1961).
19. Cornthwaite, D., and Haslam, E., unpublished results.
20. Grisebach, H., and Kellner, S., *Z. Naturforsch.*, **198**, 125 (1964).
21. Harborne, J. B., and Corner, J. J., *Biochem. J.*, **81**, 242 (1961).
22. Birch, A. J., and Donovan, F. W., *Aust. J. Chem.*, **6**, 360 (1953).
23. Neish, A. C., Underhill, E. W., and Watkin, J. E., *Can. J. Biochem. Physiol.*, **35**, 219 (1957).
24. Neish, A. C., and Watkin, J. E., *Can. J. Biochem. Physiol.*, **38**, 559 (1960).
25. Grisebach, H., *Z. Naturforsch.*, **12B**, 227, 597 (1957); **13B**, 335 (1958).
26. Grisebach, H., and Bopp, M., *Z. Naturforsch.*, **14B**, 485 (1959).
27. Grisebach, H., and Patschke, L., *Chem. Ber.*, **93**, 2326 (1960).
28. Bopp, M., and Matthis, B., *Z. Naturforsch.*, **17B**, 811 (1962).
29. Davis, B. D., "Amino Acid Metabolism", Johns Hopkins Press, Baltimore (1955).
30. Gibson, F., and Gibson, M. I., *Biochem. J.*, **90**, 248, 256 (1964).
31. Gibson, F., and Jackman, L. M., *Nature, Lond.*, **198**, 388 (1963).
32. Plieninger, H., *Angew. Chem.*, **1**, 367 (1962).
33. Plieninger, H., and Keilich, G., *Chem. Ber.*, **92**, 2897 (1959).
34. Plieninger, H., Ege, G., Fischer, R., and Hoffman, W., *Chem. Ber.*, **94**, 2106 (1961).
35. Plieninger, H., Ege, G., Fischer, R., and Hoffman, W., *Chem. Ber.*, **94**, 2106 (1961).

36. Srinavasan, P. R., Katagari, M., and Sprinson, D. B., *J. biol. Chem.*, **234**, 713 (1959).
37. McCalla, D. R., and Neish, A. C., *Can. J. Biochem. Physiol.*, **37**, 537 (1959).
38. Battersby, A. R., *Proc. chem. Soc.*, 189 (1963).
39. Freudenberg, K., *Nature, Lond.*, **183**, 1152 (1959); *J. pure appl. Chem.*, **5**, 9 (1962).
40. Freudenberg, K., Harkin, J., Reichert, M., and Fukuzumi, T., *Chem. Ber.*, **91**, 581 (1958).
41. Acerbo, S. N., Schubert, W. J., and Nord, F. F., *J. Am. chem. Soc.*, **80**, 1990 (1958).
42. Jeffs, P. W., *Proc. chem. Soc.*, 80 (1962).
43. Suhadolnik, R. J., Fischer, A. G., and Zulalian, J., *Proc. chem. Soc.*, 132 (1963).
44. Battersby, A. R., Binks, R., and Yeowell, D. A., *Proc. chem. Soc.*, 86 (1964).
45. Bendall, D. S., and Gregory, R. P. F., "Enzyme Chemistry of Phenolic Compounds", Pergamon Press, Oxford (1963).
46. Haworth, R. D., *J. chem. Soc.*, 448 (1942).
47. Grisebach, H., and Ollis, W. D., *Experentia*, **17**, 4 (1961).
48. Bunton, C. A., Kenner, G. W., Robinson, M. J. T., and Webster, B. R., *Tetrahedron*, **19**, 1007 (1963).
49. Kosuge, T., and Conn, E. E., *J. biol. Chem.*, **234**, 2133 (1959).
50. Brown, S. A., *Phytochemistry*, **2**, 137 (1963).
51. Austin, D. J., and Meyers, M. B., *Tetrahedron Lett.*, 765 (1964).
52. Robinson, R., *Nature, Lond.*, **137**, 172 (1936).
53. Seshadri, T. R., *Tetrahedron*, **6**, 169 (1959).
54. Grisebach, H., "Recent Developments in the Chemistry of Natural Phenolic Compounds", (W. D. Ollis, ed.), Pergamon Press, Oxford (1961).
55. Ollis, W. D., and Sutherland, I. O., "Recent Developments in the Chemistry of Natural Phenolic Compounds", (W. D. Ollis, ed.), Pergamon Press, Oxford (1961).
56. Roux, D. G., and Paulus, E., *Biochem. J.*, **78**, 785 (1961).
57. Roux, D. G., and Maihs, A. E., *Biochem. J.*, **78**, 12B (1961).
58. Roux, D. G., and Paulus, E., *Biochem. J.*, **82**, 324 (1962).
59. Swain, T., and Hillis, W. E., *J. Sci. Fd Agric.*, **10**, 533 (1959).
60. Hillis, W. E., "Wood Extractives", (W. E. Hillis, ed.), p. 60, Academic Press, London (1962).
61. Albrecht, H., and Bernard, K., *Helv. chim. acta*, **30**, 627 (1947).
62. Brucker, W., *Planta*, **48**, 627 (1957).
63. Brucker, W., and Hashem, M., *Flora*, **157**, 57 (1962).
64. Grisebach, H., and Vollmer, K. O., *Z. Naturforsch.*, **188**, 753 (1963).
65. El-Basyouni, S. Z., Chen, D., Ibrahim, R. K., Neish, A. C., and Towers, G. H. N., *Phytochemistry*, **3**, 485 (1964).
66. Zenk, M. H., and Müller, G., *Z. Naturforsch.*, **19B**, 398 (1964).
67. Zenk, M. H., *Z. Naturforsch.*, **19B**, 83 (1964).
68. Conn, E. E., and Swain, T., *Chemy Ind.*, 592 (1961).
69. Haslam, E., Haworth, R. D., and Knowles, P. F., *J. chem. Soc.*, 1854 (1961).
70. Gross, S. R., *J. biol. Chem.*, **233**, 1146 (1958).
71. Haslam, E., unpublished observations.
72. Schmidt, O. T., *Fortschr. Chem. Org. NatStoffe*, **14**, 71 (1956).
73. Schmidt, O. T., and Mayer, W., *Angew. Chem.*, **68**, 103 (1956).

74. Reichel, L., Haussler, R., Pastuska, and Schulz, M., *Naturwissenschaften*, **44**, 89 (1957).
75. Schmidt, O. T., Chemical Society Meeting, Sheffield (1962).
76. Mayer, W., Bachmann, R., and Kraus, F., *Chem. Ber.*, **88**, 316 (1955).
77. Schmidt, O. T., *Leder*, **14**, 40 (1963).
78. Wenkert, E., *Chemy Ind.*, 906 (1959).
79. Haslam, E., *Phytochemistry*, **4**, 495 (1965).
80. Perkin, A. G., and Uyeda, Y., *J. chem. Soc.*, 66 (1922).
81. Kutani, N., *Chem. pharm. Bull.*, *Tokyo*, **8**, 72 (1960).
82. Pojarkova, A., *Trudȳ bot. Inst. Akad. Nauk. SSSR*, **I**, 225 (1933).

General Index

A

Acacetin, 33
Acer tannin, 96, 100, 105, 106, 115, 116, 169
Acetate pathway, 60, 143, 146
Afromosin, 36
(+) Afzelechin, 47
Algarobilla, 92, 131, 133
Aluminium chloride, 38
Amaryllidaceae alkaloids, 155, 163
Amino acids, from collagen, 3, 5
Anthocyanins, 25, 26, 79
Apigenidin, 25, 44
Apigenin, 32
Arbutin, 24, 26, 94, 95, 97, 98, 145
Atrolactic acid, 45, 46
Aucuparin, 30
    methoxy, 30
Aurone, 33, 37

B

Biflavonyls, 84
Biosynthesis
    condensed tannins, 160
    coumarins, 155
    flavonoids, 157–162
    gallic acid, 162–164
    hydrolysable tannins, 125, 164–167
    lignin, 66, 67, 152, 153, 154
    phenolic alkaloids, 155, 163
    stilbenes, 61
    thyroxine, 84
Brevifolin, 100, 133–135, 166
    carboxylic acid, 100, 124, 133–135
Brevilagin I and II, 131
Butein, 29, 30
(−) Butin, 29, 30

C

Caffeic acid, 27
Caffeoylquinic acids, 93, 96
(+) Catechin, 15, 28, 37, 69, 119
    biosynthesis, 158–160
    chemical reactions, 41, 43
    chromatography, 28, 29

epimerization, 47, 48
infra-red studies, 22
isolation, 28
paper chromatography, 20, 21, 23
polymerization, 69, 70, 71, 72, 73, 74, 81–83
stereochemistry, 20, 44, 45, 46, 47
Chalcones, 19, 23, 34, 37, 44, 148, 157–159
Chebulagic acid, 100, 105, 123, 127, 130–131
Chebulic acid, 100, 105, 124, 127–131
Chebulinic acid, 100, 105, 127, 130–131
Chemotaxonomy, 167
Chestnut, 92, 136
Chinese gallotannin, 92, 100, 104–108, 110–113
Chlorellagic acid, 131–132
Chlorogenic acid, 10, 85, 102
Chorismic acid, 149–150
Chromatography
    cellulose, 28, 100, 101
    magnesol, 29, 30
    paper, 18–26, 67
    polyamide, 26, 29, 100
    silica, 29
    thin layer, 26–27
    vapour phase, 27
Chrysin, 61
Colchicine, 155
Collagen
    amino acids from, 3, 5
    electron diffraction of, 2
    fibres, 2, 3, 6, 7
    fibrils, 2, 5–8
    helical structure, 4
    tanning of, 1, 79
    tropocollagen from, 5, 6
    X-ray diffraction, 2, 3, 5
Condensed tannins
    biosynthesis, 160
    catechin hypothesis, 15
    estimation, 15
    formation, 11, 15, 68, 69
    isolation, 17, 18, 67

Botanical Index

A

Acacia sp., 29, 47, 48, 133
  catechu, 14, 15, 28, 47, 81
  harpophylla, 57
  intertexta, 50, 57
  melanoxylon, 48, 50, 57, 59
  mollisima (syn. *mearnsii*), 14, 47, 56, 57, 67, 68, 159
Acer sp., 105, 106
  campestre, 169
  ginnale, 116, 168, 169
  griseum, 169
  macrophylla, 169
  monspessulanum, 169
  palmatum, 169
  pennsylvanicum, 169
  platanoides, 168, 169
  pseudoplatanus, 169
  rotundilobum, 169
  rubrum, 169
  saccharinum, 99, 169
  saccharum, 169
  spicatum, 169
  tartaricum, 169
Aerobacter aerogenes, 148, 149
Afzelia sp. 41, 47
Anogeissus latifolia, 106
Arctostaphylos uva-ursi, 105
Aspergillus niger, 99, 102, 103

B

Bergenia sp., 94
Butea frondosa, 57

C

Caesalpinia
  brevifolia, 92, 131, 133
  coriaria, 92, 103
  spinosa, 92, 100, 106, 113
Camellia sp., 47, 106
  sinensis, 85, 119
Castanea sp., 47, 106
  sativa, 14, 92, 106
  vesca, 136

Chaemaecyparis obtusa, 84
Cleistanthus collinus, 57
Colchicum autumnale, 155
Cotinus coggyria, 57
Crataegus oxyacantha, 58, 68

D

Dalbergia sp., 31
Derris elliptica, 32

E

Escherichia coli, 148
Eucalyptus sp., 67
  astringens, 14, 59
  calophylla, 47, 57
  pilularis, 57
  wandoo, 14, 17, 59
Eugenia maire, 122
Euphorbia formosana, 122

F

Fagopyrum
  esculentum, 157, 159
  tataricum, 157

G

Gaultheria procumbens, 163
Geranium sp., 105
  pyrenaicum, 163
Gingko biloba, 84
Gleditsia japonica, 57
Guibourtia coleosperma, 57

H

Haemanthus natalensis, 155
Hamamelia sp., 106
Hamamelis virginica, 115
Hydrangea macrophylla, 156

J

Juglans regia, 138

L

Larix decidua, 14
Lavandula officinalis, 156

DATE DUE